WIRELESS

The Revolution in Personal Telecommunications

The Artech House Mobile Communications Series

For a complete listing of *The Artech House Telecommunications Library,*
turn to the back of this book.

WIRELESS

The Revolution in Personal Telecommunications

Ira Brodsky

Artech House
Boston • London

Library of Congress Cataloging-in-Publication Data
Brodsky, Ira
    Wireless: the revolution in personal telecommunications/Ira Brodsky
    p. cm.
    Includes bibliographical references and index.
    ISBN 0-89006-717-1(acid-free)
    1. Wireless communication systems. I. Title.
TK5103.2.B76   1995   95-6097
384.5–dc20   CIP

British Library Cataloguing in Publication Data
Brodsky, Ira
    Wireless: Revolution in Personal Telecommunications
    I. Title
    621.38456
    ISBN 0-89006-717-1

© 1995 ARTECH HOUSE, INC.
685 Canton Street
Norwood, MA 02062

International Standard Book Number: 0-89006-717-1
Library of Congress Catalog Card Number: 94-6097

10 9 8 7 6 5 4 3 2

▼▼▼

CONTENTS

CHAPTER 1
▼▼▼

SILICON RADIOS

1.1 THE NEXT PC

Personal communicators will be to telecommunications what the personal computer is to computing. While some people fear technology will grow increasingly distant and sterile, just the opposite has been occurring. The more computer technology has advanced, the more accessible it has become to ordinary individuals. Now wireless communications—once the exclusive realm of broadcasters, ham radio tinkerers, and private license holders—is about to become an everyday tool.

The wireless revolution promises *Star Trek*–style communicators. We will be able to make and receive phone calls from virtually anywhere using devices small enough to fit inside our ears (our own voices will be picked up via *bone conduction*). Mobile access to the public telephone network will become as common—perhaps more common—than fixed access. But *Star Trek* underestimated the power of personal communications: we will not only use tiny communicators to talk, but to send and receive text, graphics, video, and other types of information. In the 23rd century, users may hold their communicators up to their eyes rather than their mouths.

There is much more to personal communications than cheaper, better cellular telephone service. Mobility is only half of the story: the tools and services to harness mobility are the other half. As communication networks shift from wire termination–based to person-based, we will be forced to develop new techniques to filter, translate, and redirect the content of our communications. Personal communications is not about giving employers and telemarketers access to individuals anytime, anywhere; it is about giving individuals *more control* over communications.

1

The personal communications revolution will overthrow the existing telecommunications and computer industries. For decades, the threat of five different companies stringing cables along the same streets has helped justify local telephone service monopolies established and maintained by governments. Wireless communications removes this final obstacle to competition in the local loop. And unlike the desktop boat anchors peddled by today's computer industry, personal communications will create the first truly "personal" computers—computers you can take with you anywhere.

Microprocessor-based communicators will undergo the same reduction in size, weight, power consumption, and cost—and rapid performance gain—that characterized the personal computer revolution. While entrenched players will seek refuge in size (as evidenced by the flurry of mergers and acquisitions underway), the real action will take place at startup companies with no existing revenue stream to protect. Just as personal computers resulted in whole new hardware and software industries, personal communicators are about to launch a cottage industry of information brokers.

But the technological key to the personal communications revolution is not smaller devices; it is digital radio and infrared transmission technology. For the first time in history, the wireless spectrum will become readily accessible to the general population for personalized use. Digital wireless technology will transform the spectrum shortage—at least temporarily—into a spectrum glut.

Digital transmission will also enhance the performance of wireless communications. Digital processing techniques will steadily conquer noise and interference. And digital radio will overcome the biggest psychological barrier to the widespread acceptance of wireless communications: the fear that privacy will be sacrificed.

In the end, personal communications may owe its success to yet another benefit of silicon. By combining a radio, telephone, and computer in one device, and using that device over networks possessing distributed intelligence, telecommunications will undergo a fundamental change. Cliches like "communications anytime, anywhere" miss the point. The purpose of personal communications is not to make telecommunications more burdensome, it is to make it *less* burdensome. Today's knowledge workers are overwhelmed by the volume of communications to which they are expected to respond. Personal communications will enable people to offload communications to machines.

1.2 BACK TO MARCONI

The term *wireless* was popular in the early days because it described best what radio was not. Its first job was to communicate where no wires could go—with ships at sea. Although wireless communications emerged in the late 1890s, the idea of broadcasting did not materialize until World War I. Mobile communications was even more unthinkable in the beginning; not until World War II engendered a race for superior battlefield communications did mobile radio come into its own.

It is easy for us, many decades later, to lose sight of the opportunities and challenges of those early days. It was not at all clear, back then, that wireless would prove useful for communications between fixed points. For example, wireless pioneers experimented with inductive techniques for communicating across rivers and lakes, in which the "antenna" length was greater than the distance between the two parties. Many despaired of any long-term prospects for wireless due to mutual interference between unrelated users; the idea of operating on different frequencies awaited the invention of tunable radios.

As much as we try to plan the evolution of new markets, technology tends to prosper for its own reasons. Many people believed the microcomputer would succeed as a home computer—storing recipes, balancing the family checkbook, and turning on the coffee maker in the morning. Quite unexpectedly, the spreadsheet catapulted the microcomputer into the role of essential business tool. Unless we are clairvoyant, we probably do not know how personal communications will succeed, either. And perhaps it is a good thing; it is best to keep as many options open as possible during a technology's embryonic phase. Sometimes failures pave the way to future successes.

With the rise of the vacuum tube, the word *radio* came to displace *wireless*, much as *automobile* replaced *horseless carriage*. People are more receptive to familiar terms and ideas. And now *wireless* is enjoying a revival. The reason is simple: appliances long associated with tethers are becoming unplugged. Telephones, computers, fax machines, and other devices are popping up in parks and shopping malls and on sidewalks and beaches.

Eventually, *wireless* will be replaced by a more contemporary label. Just as *radio* gave modern expression to the use of radiant energy for communications, a new term will arise to convey the marriage of computers and radios—a marriage that will produce completely new forms of communication.

1.3 COMPUNICATORS?

The first computers were giant number crunchers—used for esoteric purposes like calculating the constant pi to 500 decimal places. Over the decades, the computer's mission has changed. As Apple Computer's Lawrence Tesler showed in his 1991 article in *Scientific American*, the first major shift occurred in the 1970s with the advent of time-shared access to centralized hosts. In the 1980s, the introduction of desktop computers and printers made desktop publishing the computer's most popular function. So what would be the primary task for computers in the 1990s? Tesler predicted communications, with mobility at the very center of activity.

History rarely unfolds as neatly as we would like. But by 1994, there was ample evidence that Tesler was right. Millions of desktop computer users are connected to corporate local area networks (LAN), the Internet, and other networks. While the portable versions of standard desktop computers—laptops and notebooks—are just beginning to come online, there are already millions of computers whose chief function is mobile communications. Every cellular telephone harbors a computer—complete with microprocessor, memory, and input/output (I/O)—and there are over 18 million in use in the United States alone.

Computers and radios have combined in three stages. In the first phase, radios began incorporating hybrid integrated circuits (IC) for status, identification, and emergency signaling (mid-1970s). In the second phase, marked by the introduction of microprocessor-based pagers and cellular telephones in the early 1980s, radios became full-fledged computers operating under program control. In the third phase, we have now come full circle: portable computers are starting to sprout antennas.

Perhaps the first radio to contain an (optional) microprocessor was Motorola's modular MX-300. Microprocessors enabled automatic trunking. Instead of the dedication of a radio channel to a single licensee 24 hours per day, 7 days per week, trunked radios allow a group of users to share a smaller pool of channels. For example, through switching and contention, 100 users can share 10 channels. Trunking is predicated on the fact that not every user is active at all times; the number of channels need only equal the maximum anticipated number of simultaneous users. In a properly designed trunked radio system, each user is unaware that they are sharing frequencies with others.

Prior to the introduction of computerized trunked radio, there were services that employed full-time operators to monitor and assign channels. The operator could be reached on a channel assigned for that purpose. With the introduction of the microprocessor, it became possible to design radio networks that could automatically identify users and assign channels. This development led to the formation of two entirely new industries: specialized mobile radio (two-way private radio for business and government) and cellular telephone.

Every cellular telephone may be a computer, but not every cell phone is fully digital. While microprocessors control call setup, maintenance, and teardown, most cellular phones still rely on analog transmission. Once they upgrade to digital transmission, cellular telephone and specialized mobile radio (SMR) systems will become end-to-end computer networks. Eventually, these networks will not care whether the bits represent voice or data (currently they do care). But the transition to digital will not be completed according to a plan; it can only be completed when a sufficient number of customers decide to pay for digital service.

Digital radio offers a number of theoretical advantages, including enhanced performance and reliability, improved privacy, lower cost per unit of information transmitted, and powerful end-to-end features. Because digital transmission requires digital content, it will become possible to send multimedia messages (containing a mixture of voice, text, and video) and freely convert information from one format to another (for example, from electronic mail (e-mail) to alphanumeric page). But the big payoff will be the increased capacity that will result from digital's spectrum-sharing capabilities.

Digital processing techniques can be employed to combat noise and interference, create intelligent antenna systems, and recover bits received in error. But digital radio is not superior to analog in all respects. For example, it actually requires *more* bandwidth to transmit an inherently analog signal (e.g., speech or music) as a digital bit stream. And digital radio links do not degrade as gracefully as analog. While analog voice transmissions may become harder to understand as the background noise increases, the human brain does a marvelous job compensating for the occasional lost syllable. In contrast, digital radio communications is often an all-or-nothing affair. Modest amounts of noise have little or no noticeable effect; but once the bit error rate (BER) exceeds a certain threshold, the signal completely vanishes.

One branch of digital radio—mobile data—traces its roots to packet radio techniques pioneered by amateur radio operators. Packet radio is based on packet-switching protocols developed for landline data networks. By inserting user information in data "envelopes" containing the origin and destination addresses, multiple users can share the same radio channel. Unfortunately, most packet radio systems employ narrowband channels (12.5 to 30 kHz wide), which quickly become traffic bottlenecks. Nevertheless, the development of packet radio has led to the first appearance of computers with antennas sticking out of their sides.

Simultaneously with the merging of computers and radios, we are also witnessing a blurring of the line separating business and personal life. Every business user is also a consumer—a fact not lost on computer and telecommunications marketing professionals. As employers demand longer hours and more time on the road, it becomes increasingly difficult (if not impossible) to identify a specific time at which the knowledge worker's business day comes to an end. Consequently, business users are learning to use their cell phones and pagers to stay in touch not only with their offices and customers, but with their families.

1.4 THE FIRST PERSONAL COMMUNICATORS

While this book is primarily about what lies ahead, make no mistake about it, the personal communications revolution has already started. Over the past decade we have seen explosive growth in wireless devices like residential cordless telephones, cellular telephones, pagers, and TV remote controls.

1.4.1 Cordless Phones

Although the success of residential cordless telephones is widely recognized, it is one of the most misunderstood markets. According to some telecommunications industry pundits, the deficiencies of residential cordless telephones will lead to a new market—the personal communications market. Nothing could be farther from the truth.

The success of analog cordless telephones does not, as some have suggested, prove the demand for cordless is so strong that customers have knowingly purchased unsatisfactory products. Instead, this success demonstrates the dynamics of a real consumer market—a market that is anything but monolithic. Despite the introduction of superior 900-MHz digital cordless phones, analog cordless phones continue to provide the price and performance desired by most consumers. Meanwhile, sales of 900-MHz digital cordless phones priced at $200 and above have been relatively slow. Only technology enthusiasts are willing to pay a premium for digital cordless. Of course, these early buyers are generating the revenue stream that will feed the development of more highly integrated, lower cost, and lower power-consuming digital cordless phones. A mass market for digital cordless phones wili appear not because better informed consumers will select the superior product, but because vendors will learn to manufacture digital cordless phones for less than it costs to make analog phones.

1.4.2 Cellular Telephony

Cellular telephony has emerged as one of the greatest success stories of the electronics industry. We should not forget, however, that there were many "experts" who did not anticipate cellular's success. (By some standards, cellular is still not all that successful, having achieved a mere 8% market penetration rate in the United States.) Millions of businesspeople now turn what was once wasted commuting time into productive business time. Before cellular caught on, however, many of these same people insisted they would never use a phone in their car; they valued their daily commute as a time for thinking.

Cellular telephone presents a classic case of high-tech market evolution. The first cell phones had to be installed in vehicles and sold for over $3,000 each; they were primarily adopted by people who made their money on the phone and spent considerable time in their automobiles, such as sales executives, small business owners, and investors. As cell phone prices began to decline, the market broadened to include real estate brokers and lawyers. (Unlike peak cellular air time, the price of which has declined only slightly, cell phone manufacturing is fully competitive.) But it was not until the early 1990s that a consumer market for cellular telephones emerged, largely due to skewed traffic patterns that left cellular networks practically idle during evenings and on weekends. Cellular carriers were happy to sign up consumers at low monthly rates (e.g., $30) once they were convinced consumers would have little impact on peak hour traffic.

How did cellular radio transform the car phone from an exclusive toy of the rich and famous into an everyday business tool and even a consumer item? By increasing spectrum capacity. The traditional solution to the problem of serving more users within a fixed chunk of radio spectrum had been to reduce the bandwidth allotted to each. For example, in some parts of the radio spectrum, the Federal Communications Commission (FCC) has split 25-kHz channels into twice as many 12.5-kHz channels. But you cannot get something for nothing—narrower channels carry less information and are more vulnerable to impairments like multipath fading.

No amount of slicing and dicing could turn the old mobile telephone service (MTS) (and later the improved mobile telephone service (IMTS)) into a universal mobile telephone service. That would require a larger spectrum allocation. The FCC resisted such a move until it was convinced a new technology—cellular radio—would deliver the necessary capacity gain.

Cellular radio represented a radical new approach to spectrum management. The concept—which first appeared in print in the late 1940s—is generally credited to Bell Labs. Rather than attacking the bandwidth consumed by each user individually, cellular optimizes the bandwidth consumption of a population of users over an entire metropolitan area. Instead of slicing up channels to increase capacity, cellular radio carves up geography.

The key to cellular telephone is *frequency reuse*. The service area is divided into cells, usually represented as hexagons neatly tiled on a map. (Some engineers prefer to represent the cells as slightly overlapping circles.) Frequency reuse is achieved by dividing the cells into clusters and available channels into groups, so that the number of cells per cluster is the same as the number of channel groups (typically seven). Each channel group may then be assigned to a cell within each cluster in such a way that adjacent cells never share the same channels. (There is, however, an important exception discussed in Chapter 4.)

Of course, the cellular scheme is of little benefit if signals can be heard several cells away. There are, therefore, two prerequisites for a successful cellular radio system: (1) low-power transmitters must be employed at both the base stations (<100W) and mobiles (0.6W to 10W), and (2) a frequency band must be selected that is well suited to short-range communications. The 800- and 900-MHz bands meet this requirement.

The first commercial cellular telephone network was placed in service by Ameritech in Chicago in 1983. In theory, cellular radio promised limitless capacity. A cellular network could meet the demand, no matter how high, by merely increasing the number of cells (which could be accomplished by making the cells smaller). According to the mathematics, a 50% reduction in cell radius should yield a fourfold increase in traffic capacity over a given service area.

But the solution is not as elegant in practice as it is in theory. The hardware at each cell site typically costs $300,000 to $500,000. Each cell site requires real estate and a dedicated phone line back to the network switching center—both of which may involve recurring monthly expenses. A higher cell density also puts a greater processing burden on the network: the smaller the cells, the more frequent the cell handoffs. Not surprisingly, cellular network operators are very cautious about shrinking their cells.

Since adjacent cells never use the same channels, *handoffs* are required in order to continue a conversation as the mobile user drives from cell to cell. This entails dropping the existing channel (a pair of frequencies) and resuming the call on another. As we reduce cell radius, the rate at which handoffs are required grows geometrically. The task of locating the mobile user to direct an incoming call is also made more difficult. These functions are managed by the mobile telephone switching office (MTSO).

The young cellular telephone industry also faced a number of business challenges. One was how to bill calls. In the landline model, for the most part, the originator must pay. Under the old Bell System, if a number was dialed but there was no answer, or the called line was busy, there was no charge. Cellular telephone is different. Valuable air time is consumed whether cellular subscribers originate or answer a call, so they must pay either way. And since the dialing process also consumes air time, many networks begin charging the moment the subscriber hits the "send" button (causing the cell phone to transmit the dialed number sequence to the network).

The cellular telephone network was designed under the assumption that subscribers are always in motion. A call is routinely handed off as the subscriber drives from one cell site to the next. The decision to hand off a call is made by the network alone, and is based on a series of measurements from both the current (serving) cell and surrounding cell sites. The network may also hand off calls to redistribute traffic. This is often done when a cell reaches full capacity; the network searches for calls that might be continued from an adjacent cell. (This leads to an interesting phenomenon: a handoff may occur even when the subscriber is stationary.) In the event of a handoff, the voice channel is reassigned via a data burst transmitted by the base station over the current voice channel during what is known as the *blank and burst* period (i.e., the mobile receiver's speaker is muted for approximately 200 ms while the radio is receiving the data burst).

One problem that continues to plague the U.S. cellular telephone industry is that licenses were granted on a city-by-city basis. Even the largest cellular carrier cannot provide seamless nationwide service. When users travel outside their home network, they are *roaming*. This presents three major challenges to cellular carriers: (1) processing legitimate call requests but preventing access fraud, (2) automatically redirecting incoming calls to the subscriber who is roaming, and (3) providing intersystem handoffs so that a call may be continued as the subscriber travels between adjacent networks. Solving these problems is a prerequisite to seamless personal communications.

The U.S. cellular telephone industry is well positioned to become a leading provider of personal communications services. AT&T is merging with McCaw Cellular, and Bell Atlantic Mobile Systems and NYNEX are merging their cellular operations, as are U.S. West and AirTouch Communications. Cell phones are becoming smaller and more power-efficient. Cellular carriers are upgrading to digital (see Chapter 4). There are no fewer than four different cellular mobile data solutions either under development or being introduced. The race to see who will dominate the personal communications industry has begun.

1.4.3 Paging

Cellular telephone was not the first broad-based mobile communications network. Paging networks emerged in the mid-1950s primarily as single-site systems, developing into metropolitan area networks in the 1960s. The original paging networks transmitted a sequence of analog tones (corresponding to the user's paging address) to select a specific pager. With this *alert-only* paging, users were simply notified by a beeping or vibrating pager that they had messages to be retrieved via telephone. Digital paging emerged in the late 1970s, enabling numeric paging (for transmitting the phone number of the person requesting the page) and alphanumeric paging (for sending very short text messages).

Many predicted paging would die out with the introduction of cellular telephone. Ironically, paging subsequently experienced its greatest growth. There are many reasons for this growth. Cellular phone users must pay for incoming as well as outbound calls; some cellular subscribers, therefore, use their pagers to screen incoming calls. They can decide whether to return a page from their cell phone, or wait until they have access to a wireline phone. Others have a cell phone installed in their car, but prefer to carry a pager when on foot. And some only use their cell phone in their home city, taking only the small pager with them when they go out of town.

Another reason—besides low cost—that paging networks have continued to prosper is their reliability. Because they are one-way, paging networks must successfully deliver each message on the first try. This is accomplished through simulcasting; that is, the same message is broadcast simultaneously over multiple transmitters with overlapping coverage. Simulcasting provides excellent building penetration; if the message does not get through from one direction, it usually gets through from another. The drawback to simulcasting is that it creates multipath interference, forcing paging networks to operate at low speeds.

Multipath interference occurs when signals from a transmitter take multiple paths to the receiver, and some of the copies arrive out of phase. For example, if two copies of the same signal arrive with equal strength but 180 degrees out of phase, they will completely cancel each other out (Figure 1.1). More likely, there will be multiple signal paths, with a range of attenuation and delay, and the receiver will see a combined but somewhat blurred signal. This blurring may make it difficult for the receiver to correctly identify adjacent symbols. (Each new signal from the transmitter is called a *symbol* and may represent one or more bits of data.)

There is, however, a way around this problem. By transmitting at relatively low symbol rates (typically yielding data rates of 1,200 bps or less), paging networks can minimize intersymbol interference (ISI). The lower the rate, the longer the duration of each symbol, and the less likely that differing path delays will cause consecutive symbols to completely overlap.

As the number of paging subscribers has expanded, prices have declined—taking profit margins with them. Paging operators are being forced to look for new growth opportunities. In countries such as China, where most individuals still do not have their own phone, paging has served as a somewhat primitive substitute for landline communications. Users invent numeric codes representing messages such as "please come home now." Thanks to vast new markets like China, the global paging business is booming.

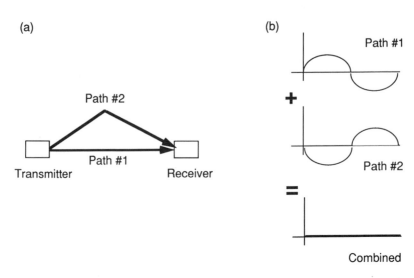

Figure 1.1 Multipath fading results when (a) signals take different paths and (b) arrive at receiver out of phase.

In the United States, the main profit growth opportunities are nationwide and enhanced paging. Nationwide networks such as SkyTel provide traveling professionals with a convenient way to remain accessible with a device small enough to be clipped to their belt or tossed in a pocket and capable of running off the same battery for a couple of months. In addition to nationwide paging, vendors are busily pursuing enhanced services such as wireless e-mail (forwarding e-mail messages to pagers), acknowledgment paging (the ability to confirm receipt of a page), answerback paging (the ability to respond to a page with selectable, preprogrammed messages), two-way paging (two-way text messaging), and even voice paging (sending brief digitized voice "clips" to specially designed pagers).

But these enhanced services will require more advanced technologies. Conventional paging networks were designed to handle a large number of very short messages. A 12,000-byte digitized voice message consumes more than 1,000 times as much airtime as a single numeric page, but one could not expect to charge 1,000 times as much per use! New high-speed protocols such as the European radio message system (ERMES) and Motorola's proprietary FLEX will increase the capacity of paging networks; ERMES operates at 6,250 bps, while FLEX runs up to 6,400 bps. By running at higher speeds, paging networks will be able to accommodate a larger number of numeric pagers or a mix of numeric and alphanumeric pagers.

SkyTel, a subsidiary of Mobile Telecommunications Technologies, (MTEL), operates the largest nationwide paging network in the United States and is a pioneer in the field of computer messaging, providing services such as SkyPager and Sky-Word. Using synthesized voice prompts, SkyPager walks callers through sending a page (usually a number to call back). Each page is sent to an earth station in Palo Alto, California, where it is relayed to the Westar IV communications satellite and broadcast to each of SkyTel's terrestrial transmitter sites. All of SkyTel's terrestrial transmitters operate on the same frequency in the 931-MHz band.

SkyWord supports transmission of messages containing 40 to 80 alphanumeric characters. Messages are input by computer or e-mail along with the SkyWord user's personal identification number (PIN). SkyWord works with the Motorola Advisor and Bravo Alpha, and NEC's IDP 7000 pager at 2,400 bps. SkyTel has also teamed up with AT&T EasyLink Services to provide an integrated wireless e-mail solution (Figure 1.2).

Motorola's EMBARC is another interesting player in the nationwide paging arena. EMBARC (Electronic Mail Broadcast to a Roaming Computer) was launched on 20 July 1992 in the top 112 markets in the United States and Canada as a one-way wireless e-mail service. The network broadcasts on 931.9125 MHz at 1,200 bps using the popular POCSAG (Post Office Code Standardization Advisory Group) paging format. POCSAG, as its name implies, was originally developed at the instigation of the British Post Office.

EMBARC provides delivery of messages up to 1,500 characters in length to portable devices equipped with Motorola's NewsStream or NewsCard radio modem. EMBARC has been attempting to open up new applications for paging such as one-to-many message delivery, updating inventory and price lists, delivering time-critical medical laboratory test results, and transmitting daily news filtered per individual subscriber profiles (through an information refinery such as that provided by Individual, Inc.).

Interestingly, EMBARC was positioned to compete with overnight couriers for

Figure 1.2 SkyTel's SkyWord Service delivers text messages, electronic mail, and news to mobile users.

the delivery of short documents. While a nationwide courier can only promise delivery by the next morning, EMBARC offered delivery within as little as 15 minutes. Couriers must base their charges on the number of destinations even if a customer is sending identical packages, whereas EMBARC charges only for message length and priority. In other words, with EMBARC one could send a new price list to 500 sales representatives scattered around the country for just a few dollars. EMBARC also developed a unique pricing strategy: instead of charging paging subscribers for each use, the company negotiated agreements with e-mail networks that required the sender (on the landline side) to pay a premium for messages handed off to EMBARC.

EMBARC's wireless e-mail service, however, enjoyed little success. The decision to use the Consultative Committee for International Telegraph and Telephone (CCITT) X.400 standard for interconnection with other e-mail services made addressing unnecessarily complicated. There was also a "chicken and egg" problem: there had to be a large number of EMBARC subscribers before e-mail users would start paying extra to reach them. As a result, EMBARC has announced a completely new service: for a flat monthly fee, subscribers can receive up-to-the-minute sports scores on a dedicated pager. They may be on to something.

While networks like SkyTel and EMBARC can increase capacity by migrating to higher speeds, communicating in only one direction is still their biggest limitation. Simulcasting—whether in a city, region, or nationwide—means transmitting the same information over multiple transmitters. If they could just identify the transmitter closest to the intended recipient, the message could be broadcast from that transmitter alone, freeing up the neighboring transmitters. Two-way paging networks, therefore, would not only allow mobile subscribers to respond to pages, they would provide vastly greater capacity. Protocols like Motorola's reFLEX promise throughputs up to 25,600 bps (combining four 6,400 bps channels) on a 50-kHz-wide forward channel (network-to-mobile subscriber), and up to 9,600 bps on a 12.5-kHz-wide response channel (mobile subscriber-to-network).

SkyTel's parent organization, MTEL, petitioned the FCC in late 1991 to establish an ambitious new service for its nationwide wireless network (NWN). NWN is a two-way packet radio network that runs 24,000 bps over a single frequency for transmit and receive. MTEL believes the conventional mobile data network architecture does not provide adequate coverage or support for low-power subscriber devices. MTEL's proposal involves two fundamental advances: (1) multitone simulcasting, and (2) dynamic zoning. Multitone simulcasting enables faster transmission. Dynamic zoning improves network efficiency by first simulcasting a "wakeup call" to the subscriber device; once the subscriber device responds, messages may be broadcast locally.

At the time MTEL submitted its petition to the FCC, most licenses were awarded through comparative hearings or lotteries. MTEL was granted the first Pioneers' Preference award, which was supposed to guarantee the firm would receive one of the first licenses for this new service. But the FCC decided to fold the advanced paging service into its *personal communications service* (PCS) rulemaking; advanced paging became known as *narrowband PCS*. It was also decided that narrowband PCS licenses would be distributed through competitive bidding (auctions). Instead of receiving a free license as originally promised, MTEL was simply given the opportunity to purchase a license at a discount calculated against the lowest winning bid.

MTEL has established a new subsidiary named Destineer Corporation to build and operate its two-way network. Applications envisioned include acknowledgment paging, two-way data messaging, fixed-location data distribution (e.g., point of sale), remote data collection, remote access to information services, wireless e-mail, and industrial remote control (e.g., remote alarm monitoring).

Another paging operator—the largest paging network in the United States (over 3 million subscribers)—is headed back to the future. In conjunction with Motorola, Paging Network (PageNet) plans to launch what it considers the "next generation of personal communications devices": pocket answering machines. PageNet's VoiceNow™ service will allow users to play (and replay) messages in the caller's own voice for what the firm claims will be a fraction of the cost of cellular telephone service.

The subscriber device is actually a two-way radio, enabling the network to locate the user before sending a complete digitized voice message from the nearest base station. A beep or vibrator will alert the subscriber that he or she has received a message.

1.5 WIRELESS ELECTRONIC MAIL

Not only is personal communications inspiring existing players to develop bold new products and services, it is also attracting a growing number of entrepreneurs. It is probably too early to say who will become the Bill Gates of personal communications, but there are some interesting candidates.

One is Geoffrey S. Goodfellow, Chairman of RadioMail Corporation. Choosing to pursue his interests in data networking and security rather than complete high school, Goodfellow worked his way up to Senior Member of Network Staff at Stanford Research Institute's Computer Science Lab. Goodfellow's interest in interconnecting wireless and wireful networks started when he was introduced to AlohaNet, the world's first packet radio network, at the University of Hawaii in 1974. Then, as Goodfellow tells it, "I wanted to be able to read and answer my e-mail while laying on the beach . . . In 1981, I devised a system to automatically forward e-mail to a MetaGram alphanumeric pager. All I wanted was to be a wireless e-mail user, but there was nothing commercially available, so I had to create my own solution."

Today, RadioMail Corporation operates a wireless two-way gateway that interconnects users equipped with portable radio modems (for use over ARDIS's or

Figure 1.3 RadioMail supports two-way mobile messaging over a variety of portable computers.

RAM Mobile Data's nationwide mobile data networks) to a variety of landline networks, including corporate e-mail systems such as Lotus cc:Mail, public systems such as MCI Mail, ATT Mail and CompuServe, and the worldwide Internet and UUCP/USENET service (Figure 1.3).

The mobile user can exchange messages with other RadioMail users or virtually any e-mail service subscriber. While RadioMail provides near instant communications between wireless subscribers, it is still subject to the delays inherent in landline e-mail. For example, most desktop users are not notified every time they receive a message. Still, RadioMail subscribers can make themselves instantly accessible via e-mail or RadioMail's operator-based messaging service. And Radio-Mail's fax gateway permits the mobile user to compose a text message that may then be sent to any fax machine anywhere in the world. With its urgent-looking cover sheet, a RadioFax™ is often hand-carried from the fax machine to its intended recipient.

1.6 BEGINNING

The personal communications revolution is not about a specific product or service, it is about a new communications paradigm. And it has already started: there are over 18 million cellular telephone users, 16 million pager users, and 40 million cordless telephone users in the United States. And there is more to personal communications than becoming untethered; it is about access to people and information with unprecedented convenience and control.

We are quickly progressing from the era of spectrum shortage to the age of spectrum glut. Plentiful spectrum will mean more than just plentiful access to the same old services. A new industry is emerging, but we cannot know in advance all of its ramifications. We only know that it will surprise us, and that such surprises tend to be pleasant.

Because the personal communications revolution is being enabled by digital radio, our point of departure is the computer rather than the radio. In the following chapters we will explore topics from wireless office telephone systems to mobile satellite services. Keep in mind that each of these "wireless computers" is subject to the rapid advance in performance and steep decline in cost that have made wired computers such an important part of our lives.

SELECT BIBLIOGRAPHY

[1]   Aitken, H. G. J., *Syntony and Spark: The Origins of Radio*, Princeton, NJ: Princeton University Press, 1985.

[2]   Boucher, N. J., *The Paging Technology Handbook*, Mendocino, CA: Quantum Publishing, 1992.

[3]   Brodsky, I., "Advanced Paging Services," *Wireless Industry Prospectus*, Vol. 1, No. 4, November 1992.

[4]   Brodsky, I., "Interview: Paul Saffo," *Wireless Industry Prospectus*, Vol. 2, No. 4, March 1993.

[5]   Brodsky, I., "Interview: Geoffrey S. Goodfellow," *Wireless Industry Prospectus*, Vol. 1, No. 4, November 1992.

[6]   Clarke, A. C., *How The World Was One*, Bantam Books, 1992.

[7]   "Motorola Wireless Communications—Milestones," research courtesy of the Motorola Museum of Electronics, Schaumburg, IL, 1994.

[8]   Tesler, L. G., "Networked Computing in the 1990s," *Scientific American*, Vol. 265, No. 3, September 1991.

CHAPTER 2

THE POLITICS OF RADIO

2.1 THE BATTLE FOR SPECTRUM

New mobile communications technologies face an enormous obstacle in the United States. Virtually all of the spectrum believed to be best suited for mobile communications—frequencies from 500 to 3,000 MHz—has already been allocated. Identifying, establishing, and distributing spectrum for the new PCS has become a major political football.

But the spectrum shortage is a manufactured problem. For example, the UHF-TV band comprises 336 MHz of bandwidth. Few cities have more than a half dozen active UHF-TV stations (each occupying just 6 MHz). The rest of the UHF-TV band has been sitting around for years collecting electromagnetic dust. Broadcasters argue that hoarding UHF-TV channels is necessary to ensure future capacity for high-definition television (HDTV). Meanwhile, the bandwidth available for television has grown by leaps and bounds. The cable TV industry is on the threshold of offering viewers a choice of 500 channels; additional channels will be delivered via new direct broadcast satellite (DBS) services. A more plausible explanation of why the National Association of Broadcasters (NAB) is reluctant to give up "its" spectrum: they are holding out for a better deal.

Claims of spectrum scarcity are not new. The NAB first warned of an impending shortage in 1925. At that time, 2 MHz was believed to be the highest usable frequency. Everything above it was given to amateur radio operators, who soon discovered the utility of higher frequencies. The upper boundary of the usable radio spectrum has risen steadily since then. Consumer products are now feasible up to about 3 GHz; industrial products up to about 40 GHz. Now, every time we raise the ceiling another gigahertz we gain an additional *1,000 MHz* of usable spectrum. So much for pessimism regarding frequencies above 2 MHz!

History suggests a combination of technology and entrepreneurial activity will conquer almost any shortage. Over the last two decades, we have witnessed impressive growth in spectrum capacity. First trunking and then space-division multiplexing (cellular) arrived to boost the capacity of narrowband radio systems. Now digital radio techniques like time-division multiplexing and code-division multiplexing promise even more capacity. We have just begun to explore the possibilities of micro- and picocellular radio networks. For the near term, we face not so much the threat of a spectrum shortage, but a spectrum glut.

New technologies call into question traditional methods of distributing spectrum access rights. Until recently, transmitting was a privilege granted in the form of exclusive licenses. License holders are permitted to transmit on specific frequencies at specific locations or—in the case of private land mobile radio licensees—over a restricted geographical area. (Sometimes, secondary licenses are granted on a noninterfering basis; that is, the secondary licensee may use the same frequency in the same or a nearby location, but must not cause interference to, and must accept interference from, the primary licensee.) The union of computers and radios and the growing preponderance of handheld communicators may render exclusive licenses obsolete.

While narrowband trunked radios enable dozens of users to contend for a smaller pool of channels, new broadband radios may permit users to share vast stretches of spectrum—what futurist George Gilder likens to a public right of way. Microprocessor-based radios will avoid interfering with each other through strict adherence to a communications etiquette similar to the one used over Ethernet LANs. Anyone will be able to access the spectrum as long as the user's equipment conforms to a minimum set of rules, so there is no need for handing out, drawing for, or auctioning off a finite number of exclusive licenses (serving as spectrum "property" rights). For example, radios might be required to hunt for and find a vacant channel (dynamic channel allocation) before transmitting. In most data applications, it is possible to take turns: radios would listen first to make sure the channel was free, restrict the duration of their transmissions, and adhere to fair rules for minimizing the incidence of and resolving collisions (i.e., two or more devices accidentally transmitting at the exact same time and interfering with each other). Radios might also be required to transmit at the minimum necessary power, use electronically steerable antennas, or employ radio technologies that facilitate spectrum sharing such as spread spectrum (see Chapter 3).

Unlicensed access to the radio spectrum is flourishing. The sale and use of cordless phones for the home, garage door openers, baby monitors, and other unlicensed devices are permitted provided the manufacturer meets certification requirements (either operation at very low power levels or the use of spread spectrum transmission). And although cellular telephone network operators are licensed, individual subscribers need not obtain licenses, as long as they use type-approved radios.

Ready access to the radio spectrum is one of the prerequisites for personal communications. Licensing the airwaves has been, at best, a necessary evil. While licensing newspapers and magazines would be recognized as a potential form of censorship, licensing radio transmitters has been thought to be the only rational means of preventing interference—the occurrence of which may lead to the same consequences. But this "pound of cure" was conceived prior to the development of "ounce of prevention" technologies—technologies that have not only increased spectrum capacity but have off-loaded most of the growth in TV broadcasting back to cables. More importantly, intelligent radios have emerged that can be programmed to avoid giving and receiving interference. Software that momentarily halts radio transmissions is infinitely fairer than a licensing process that sometimes permanently denies access.

The government must not stand between entrepreneurs and end users. The technologies that are opening up the radio spectrum to a broader market are at the same time undermining all of the old arguments favoring licensing. Only network operators should be required to obtain licenses—and then only on a nonexclusive basis. As technology continues to increase spectrum capacity, the number of service providers (each adhering to a minimum set of rules) will be limited by the demand for their services—and nothing else.

2.2 THE FCC: IN OVER ITS HEAD?

The use of the radio spectrum in the United States (with the exception of federal government use) is managed by the FCC, which is headquartered in Washington, D.C. The FCC was set up as an independent regulatory body under the Communications Act of 1934—at a time when communications consisted solely of telephone, telegraph, and radio. The FCC allocates spectrum for specific purposes, defines rules governing each service, grants licenses to users, certifies products for use over the airwaves, and polices the radio spectrum. FCC Rules are published by the U.S. Government Printing Office and are divided into parts 0 to 100 (not all currently in use). In addition to describing spectrum allocations and licensing procedures, FCC Rules define eligibility requirements for various services and procedures for submitting petitions to amend existing rules or create new ones (Table 2.1).

What is not revealed by the above description is the fact that the FCC has undergone significant changes over the past 15 years or so. Up until the late 1970s, the FCC was primarily an administrative agency. Its main tasks were processing license applications and enforcing rules. The old FCC could be described as "radio police."

Then powerful microprocessors began to invade radios. Radios with computers inside made more efficient use of radio spectrum, clearing the way for hordes of

Table 2.1
FCC Rules Concerning Mobile Communications

Part	Title
15	Radio Frequency Devices
18	Industrial, Scientific, and Medical Equipment
22	Public Mobile Service
80	Stations in the Maritime Services
87	Aviation Services
90	Private Land Mobile Radio Services
95	Personal Radio Services

new users. More importantly, computer radios opened the door to myriad new applications, and industry began seeking new spectrum to accommodate these uses. Rules that had stood for decades were suddenly obsolete.

For better or worse, the FCC's job has changed. It has been forced to off-load or abandon administrative tasks such as frequency coordination, conducting amateur radio license examinations, and awarding licenses through comparative hearings. Today, the FCC's main job would appear to be ensuring spectrum is allocated to meet the needs of evolving industries while monitoring and protecting the legitimate interests of existing users.

The new FCC can claim several successes. The reallocation of spectrum for cellular telephone and specialized mobile radio gave birth to whole new industries. Many believe these industries were *created* by the FCC. (An alternative explanation is that these industries arose when the FCC finally took down the barriers to enabling technologies.) Now many believe the FCC is once again commencing a gold rush—the PCS gold rush. Only this time they are selling the franchises (spectrum auctions) rather than giving them away (lotteries).

The FCC has clearly been thrust into a role well beyond what was intended when it was founded. At best, it muddled its way through the establishment of the cellular telephone and SMR services. But things were easier when the only combatants were the NAB and the phone company, and no one was sure cellular telephone would even work. PCS is going to be much tougher. Every special interest group is converging on Washington to get its share of this "sure thing." Congress is busy second-guessing FCC actions, and the Clinton administration is hailing PCS as a source of budget-balancing revenue, new jobs, and American leadership in global markets—all before the first commercial PCS network has been switched on. In reallocating spectrum for new services, the FCC finds itself squarely in the middle of the industrial policy-making game.

Let us take a closer look at the spectrum allocation process. Basically, there are three ways to obtain spectrum for new technologies: (1) identify and secure unassigned or reserved spectrum, (2) devise a method of sharing spectrum with an existing service, or (3) get the FCC to reallocate spectrum. The first option will not work—there is virtually no unassigned spectrum below 3 GHz (the 930- to 931-MHz advanced paging reserve, renamed *narrowband PCS*, was until recently an important exception). While spread spectrum and other techniques suggest opportunities for spectrum sharing, the details have not been worked out, and incumbent users exhibit fierce resistance at just the mention of sharing. Clearly, large tracts of spectrum can only be opened up to new technologies through *refarming* (i.e., updating technical standards) or reallocation.

The foundation of today's spectrum distribution process is the block allocation system. A block is a contiguous band of frequencies dedicated to a particular service—usually governed by a unique set of technical standards and eligibility requirements. The advantages of block allocations are that they facilitate frequency-use coordination, make it easier for manufacturers to develop products for specific markets, and enable the government to reserve bandwidth for socially desirable uses.

But block allocations have their disadvantages. The biggest problem is that once a block is assigned to a particular service, it becomes extremely difficult to change or reclaim. For example, if an assigned block is defined as consisting of 30-kHz channels, it is very difficult to get users and manufacturers to migrate to 15-kHz channels—even though the technology has advanced to the point where the narrower, more spectrally efficient channels provide the same quality of service. Nevertheless, such spectrum refarming efforts are under way.

Another problem is that once a block has been assigned, it is difficult to enlarge or reduce its bandwidth, or relocate users to another band. The result has been that some blocks are operating near capacity, while others are operating way below capacity. In some cases, the FCC has simply failed to anticipate growth.

One way to even out discrepancies between different spectrum blocks is to define them more broadly. In fact, spectral efficiency would be improved just by reducing the number of different block allocations. (Some point out that, alternatively, blocks can be defined not only by frequency, but by criteria such as time or geography.) Along these lines, the NTIA (see Section 2.3) has argued that the FCC should relax technical standards so that vendors and users may more readily exploit new technologies.

New spectrum allocations or spectrum refarming require changes in the FCC Rules. Such changes may be initiated by the FCC or by a "Petition For Rulemaking" from any source. If the FCC determines that a petition from the public is worthy of consideration, it issues a "Request For Comment" and "Reply Comments." Any interested party may comment on the petition's merit. The FCC then issues a "Notice of Proposed Rulemaking" (NPRM) or "Notice of Inquiry" (NOI). Again, comments are solicited from the public. Finally, the FCC either issues a "Report and Order" accepting the proposed rule changes, or a "Memorandum Opinion and Order" declining the request. See Figure 2.1.

The Petition For Rulemaking has been a vital tool for entrepreneurs working to introduce new technologies into the radio spectrum. Unfortunately, the process described above can (and often does) take years to complete. Many smaller firms are unable to endure the time and expense. Until recently, those who submitted a petition were forced to place their ideas in the public domain with no guarantee that

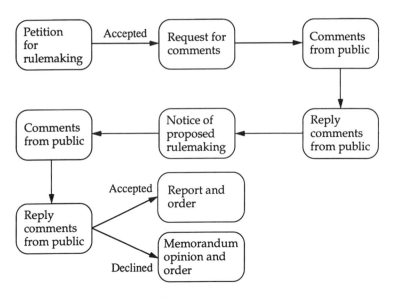

Figure 2.1 FCC rulemaking process (typical).

they—and not someone else—would be granted the first license. The "Pioneers' Preference Rule" was devised to redress this imbalance by granting the right to a first license to innovators who disclose their ideas in a petition to the FCC.

But the Pioneers' Preference has led to two major problems. First, determining who has advanced the state of the art is a subjective matter. The first Pioneers' Preference awards have been vigorously challenged. And with the introduction of competitive bidding for licenses, some argue the new Pioneers' Preference Rule is already obsolete. For example, members of Congress have voiced concern that Pioneers' Preference award winners would grab up the most valuable licenses—licenses that otherwise would have commanded handsome bids in the spectrum auctions. Having been vague about the meaning of a Pioneers' Preference from the start, the FCC now seems comfortable offering award recipients what amounts to a price discount in the auctions.

2.3 THE NTIA: NOT TERRIBLY IMPORTANT AGENCY?

Less well known is the National Telecommunications and Information Administration (NTIA)—the branch of the Department of Commerce that oversees all federal government use of the radio spectrum. There is some coordination between the NTIA and FCC, however, because a large portion of the radio spectrum is shared between government and nongovernment users.

Ostensibly to protect national security interests, the federal government has not disclosed which frequencies it uses and where. This has, to say the least, complicated spectrum sharing. Nongovernment users who want access to frequencies shared with the government must participate in a cat and mouse game: the only way to determine if a frequency is being used by the federal government is to apply for it and wait to see if the application is accepted or denied.

The NTIA also serves the executive branch in formulating and articulating telecommunications policy by conducting studies and publishing reports on various topics. Under the Bush administration, the NTIA published a controversial report entitled "U.S. Spectrum Management Policy: Agenda For the Future," which recommended, among other things, the use of spectrum auctions.

But what the NTIA envisions is a tightly managed competitive bidding process. It wants to establish minimum bidders' qualifications, restrict certain groups from participating, and give special discounts to others. If one believes, as the NTIA apparently does, that market mechanisms "are not a panacea . . .," then why bother to emulate them? Not surprisingly, the managed auctions advocated by the NTIA under the Bush administration have turned into just another source of government revenue under the Clinton administration. Instead of injecting market forces into spectrum distribution, the politicians have simply created a new (but voluntary) tax.

2.4 LICENSING: AN IDEA WHOSE TIME HAS COME AND GONE

Licensing operators of radio transmitters was conceived as a means of preventing harmful interference. New technologies, however, challenge the need for licensing. For example, individual cellular telephone users are not licensed; instead, their microprocessor-based cell phones are automatically directed to unoccupied channels. But a more likely model for the future was contained in Apple Computer's Data-PCS proposal. Apple describes an environment in which each user device adheres to a "listen before talk" radio etiquette—a sort of Ethernet over the airwaves. Instead of licensing as the first line of defense against harmful interference, Apple's proposal relies on *technology*.

Apple's Data-PCS proposal extended the concept of unlicensed operation as first defined in FCC Rules governing the use of "radio frequency devices." FCC Rules Part 15 were designed to permit the operation of intentional, unintentional, and incidental radiators without individual user licenses. *Incidental* radiators are devices such as dc motors and jewelry cleaners that may produce radio signals, but were not designed to do so. *Unintentional* radiators are devices, like personal computers and microwave ovens, that produce radio signals for purposes other than communications. It was deemed logical, therefore, that unlicensed individual use be extended to intentional radiators as long as they operate within the same restrictions that apply to incidental and unintentional radiators.

A new industry has emerged to exploit Part 15 Rules. Users and vendors want quick, convenient, and free access to the radio spectrum. While licensed services favor entrenched players who know how to navigate government bureaucracies, unlicensed services favor entrepreneurs and users on low budgets. Today, residential cordless telephones, wireless private branch exchanges (PBX), garage door openers, baby's room monitors, and wireless LANs are just some of the devices being marketed for use under Part 15.

But Part 15 is not without its drawbacks. Operation of Part 15 devices is permitted on a secondary basis only. That means they must not cause harmful interference with higher priority users, and are not protected against interference received. In other words, if a Part 15 device causes interference with a licensed user, the Part 15 user may be forced to cease operation. Furthermore, most Part 15 devices operate in the industrial, scientific, and medical (ISM) bands, where they are subordinate to retail antitheft tags, automatic vehicle monitoring systems, and even amateur radio operators. Buyers of Part 15 devices, particularly wireless LAN and PBX systems, beware!

While the use of Part 15 devices is unlicensed, manufacturers must have each product design certified in accordance with FCC Rules Part 2, Subpart J. Class A certification permits the device to be used in industrial settings. More stringent Class B requirements certify a product for use in residential environments. Certainly one can say that devices meeting Part 15 certification requirements are unlikely to cause harmful interference, and reports of Part 15 users being forced to cease operation are extremely rare.

Another interesting aspect of Part 15 is the special status granted spread spectrum devices in section 247 (see Chapter 3). While Part 15 devices are normally restricted to very low transmit power levels (e.g., 0.75 mW), spread spectrum devices are permitted to operate with up to 1W of output power. The logic behind this relatively high power level is that spread spectrum devices are inherently less likely to cause interference. This section of the Rules has been a godsend to vendors, who can produce higher performance unlicensed devices by simply using spread spectrum.

But most manufacturers of unlicensed products for business feel that Part 15 Rules are not a satisfactory long-term solution. They feel that unlicensed devices require their own exclusive spectrum. In other words, unlicensed products will become much more successful when relieved of having to coexist on the same frequencies with users that have higher priority (and often higher transmit power). This was the role conceived for the unlicensed personal communications service (UPCS).

The key to wider use of unlicensed devices is the further development and refinement of new technologies. In microcellular networks, devices can reuse frequencies at relatively short distances without interfering with one another. Spread spectrum transmitters can operate on top of the same frequencies as conventional radios, in some cases producing a barely noticeable increase in the background noise level. And frequency-agile radios operating over broadband channels can hunt for and grab momentarily clear frequencies on an as-needed basis.

2.5 REGULATION OF RADIO NETWORKS

Another issue before the FCC has been how to treat different types of mobile communications networks. For the most part, the landline world is neatly divided into private networks and common carriers. Common carriers use public right of ways and are regulated. In exchange for *universal service* (serving anyone who asks and can pay) and charging reasonable rates (often subject to approval by states' public utility commissions (PUC)), a common carrier is usually granted some form of protected market. Private networks use private facilities or lease facilities from common carriers, and are not themselves subject to regulation.

Things are not as neatly structured in the world of mobile communications. Radio waves do not know where public spaces end and private premises begin. In the wired world, communities have accepted local monopolies to avoid the disruption and unesthetic appearance of crisscrossing wires from multiple vendors. In the mobile world, there are wires, but they do not extend out to each and every user. Until recently, the FCC recognized three different types of wireless networks: private, shared private, and common carrier.

Private radio networks are built and operated for private use. A well-known example is the nationwide mobile data network used by Federal Express in its package tracking application. Shared private networks are a more ambiguous category; they may include networks that serve external customers and operate for profit. Not surprisingly, cellular telephone networks and certain paging operators (radio common carriers (RCC)) are considered common carriers.

Ostensibly to streamline regulations and eliminate inconsistencies, the FCC has restructured mobile communications into just two categories: private mobile radio service (PMRS) and commercial mobile radio service (CMRS). CMRS carriers are entities that (1) provide service for profit, (2) are connected to the public switched telephone network (PSTN), and (3) are available to a significant portion of the population. Under this definition, licensed PCS, paging, cellular, SMR, and wireless data networks are all considered commercial and are subject to common carrier regulation. But the FCC has also exempted CMRS from certain requirements of Title II of the Communications Act of 1934. The FCC will not require CMRS carriers to file tariffs and will not approve market entry or exit. But it will promote universal service and (ostensibly) protect the public from high prices. The FCC also plans to continue monitoring the competitiveness of cellular telephone, and will require local exchange carriers (LEC) to provide all CMRS carriers with reasonable and fair interconnection to the PSTN.

This change, however, arguably leads to more rather than less regulation. Merely reducing the number of categories of carriers from three to two does not automatically lessen regulation. What is more important is the fact that previously unregulated carriers, such as SMR operators, are now being lumped together with cellular telephone operators. Although the category was confusing, *private shared networks* were commercial carriers who were treated just like private networks. For the most part, they were left alone.

The FCC's track record in regulating mobile communications common carriers has proven hardly more enlightened than its regulation of landline carriers. The FCC's establishment of a cellular telephone "duopoly" has been, in terms of promoting competition, a fiasco. As often happens with regulation, the industry learned to circumvent the FCC's decision to license a *wireline* (i.e., reserved for a subsidiary of the local telephone company) and *nonwireline* (i.e., reserved for anyone but the local telephone company) cellular telephone carrier in each metropolitan statistical area (MSA). For example, in Chicago there are cellular systems owned and operated by Ameritech (the local wireline carrier) and Southwestern Bell (the local wireline carrier in St. Louis); the same two companies hold both of the cellular licenses for St. Louis in, of course, the reverse roles. Here's what the FCC sees as competition: instead of one monopoly controlling all of one market, two monopolies own one-half each of two markets!

2.6 PRIVACY, SECURITY, AND ACCESS FRAUD

Wireless technology is plagued by negative perceptions regarding security. While vendors are quick to dismiss such concerns, customers are not. A digital radio network may be as secure as any analog wireline network. But that will do little good if people believe it is inherently insecure.

In particular, people tend to associate all varieties of radio with broadcasting. There is an uneasy feeling that conversations or data sent over the airwaves may be easily picked off by users equipped with scanners purchased from mail-order dealers. As if to acknowledge this state of affairs, the U.S. government has made it illegal for anyone to eavesdrop on cellular telephone calls. (Ironically, it is perfectly legal to listen in on cordless telephone calls.)

Of course, it is not so easy to regularly eavesdrop on an identified cellular telephone user. Due to the low power of mobile/portable cell phones, one would have to follow that person around. Because cellular telephone operates full duplex over two channels, eavesdroppers need two receivers and must be able to identify and switch to the appropriate frequencies in the event of a cell handoff.

There are also systems for scrambling analog cellular phone calls. Normally, compatible scrambler/descrambler devices would be required at each end of the link. This requirement may be circumvented by routing all calls through a PBX equipped with a scrambler/descrambler. In this way, the radio leg of the communications link is always protected, even during conversations with landline users equipped with ordinary phones. In some cases, the cellular subscriber may answer a direct (un-scrambled) call, and then dial back the caller through the PBX when it is determined that privacy is required, such as a conversation between a lawyer and his or her client.

Privacy is greatly enhanced with digital radio. First, digital radio transmissions are unintelligible to mail-order scanners. Second, since both voice and data transmissions are already being transported as bit streams, they are easily encrypted. Sure, it is still possible for a determined eavesdropper with a sufficient budget to gain access to private conversations over a wireless network—but at some point it becomes easier to attack the landline side of the call.

The ascent of relatively bullet-proof digital radio technology has contributed to the U.S. government's plan to force the domestic telecommunications industry to adopt government-defined encryption. The notorious Clipper Chip and associated SkipJack encryption algorithm would enable law enforcement authorities, with proper authorization, to eavesdrop on digital telecommunications. The Clipper Chip uses a *key escrow* system for which the government would possess the master key. Federal authorities also want telecommunication networks to design in "back doors" for access by security agencies. (In other words, networks would be responsible for providing the government with the hardware ports and software protocols to get what they are looking for.) The problem with the Clipper Chip strategy, of course, is that it will not work: criminals and terrorists are not going to sheepishly use encryption for which the U.S. government holds the keys. Legally restricting domestic sales to equipment incorporating Clipper can only lead to a black market for products implementing proprietary encryption techniques.

So far, it is access fraud rather than privacy that has been the most serious mobile communications security problem. It is estimated that as much as 15% of billable cellular telephone airtime is lost to fraud. Because cellular telephones are computers, they are susceptible to tampering by non-law-abiding hackers. It is relatively easy to reprogram a cell phone to assume the identity of another legitimate user. Not surprisingly, phones reprogrammed in this manner have been used to sell international phone calls at greatly reduced rates. Perpetrators of access fraud reportedly set up shop in a parking lot or forest preserve at a prearranged time and date.

Cellular carriers have a number of weapons at their disposal to combat fraud. They can offer to block international calls for subscribers who do not anticipate any need to make such calls from their cell phone. The network can also flag suspicious calling patterns; for example, the cellular phone that suddenly places a series of international calls.

Other clues pointing to possible fraudulent use include when a cellular phone appears to be active in more than one location at the same time. Although carriers can disable cell phone service before the next call attempt, illicit hackers have responded by creating phones that "tumble" from one false identity to another. The ability to identify fraudulent activity is delayed when the illegal phone appears as a roamer (i.e., on a network other than its home network); information must be exchanged between the two cellular carriers. The migration to the more advanced signaling system no. 7 will enable cellular carriers to more quickly verify the legitimacy and good credit of a roamer.

The issue of access fraud is, understandably, a major concern for emerging personal communications. The upgrade to digital radio transmission, however, will facilitate the introduction of more powerful methods of combating access fraud. Among these is the *challenge-response* technique. The network may challenge a subscriber device at any time (e.g., during a call or even while dialing) to prove it is who it claims to be. Essentially, the network challenges the subscriber device by transmitting a number; the subscriber device performs a calculation on this number and responds by transmitting the results. Only the network and the authentic subscriber device know the correct answer. (Through the use of password-protected *smart cards*, legitimate subscribers could even use rented or borrowed handsets.)

But wireless security concerns are not limited to privacy and access fraud. There is fear that wireless data networks (especially wireless LANs) could prove vulnerable to attack by airborne computer viruses. In a wireless environment, it is not necessary to have access to cabling, floppy disks, or even the inside of the building to plant a virus capable of damaging files and crashing systems.

While radio interference is an annoyance to consumers, it could have more serious consequences in business environments. For example, there have been reports of portable radios interfering with sensitive patient monitoring devices in hospitals. *The Wall Street Journal* (June 15, 1994), reported a pacemaker was unnecessarily implanted in a patient after radio emissions from a nearby TV blanked out portions of an electrocardiogram trace.

2.7 WILL POTENTIAL HEALTH HAZARDS SCRUB THE PERSONAL COMMUNICATIONS REVOLUTION?

Are low-power transmitters dangerous to your health? The dilemma faced by the emerging personal communications industry is that it is impossible to prove a negative. But that is precisely what is being demanded.

An uproar was created when the spouse of a cellular telephone user and then an engineer working for Motorola filed lawsuits claiming the use of handheld cellular telephones caused brain tumors. However, if one looks at the number of brain tumors reported to date by handheld cell phone users and then the total number of handheld cell phone users, it can be seen that the incidence of brain tumors among handheld cell phone users is lower than the incidence of such tumors among the general population. On the basis of that information, one would be forced to conclude that the use of a handheld cellular phone serves as a prophylactic against brain cancer!

Nevertheless, it must be noted that the amount of radiation incurred is inversely proportional to the square of the distance from the radiating antenna. At 3 feet away, the radiation received by the brain from a cellular telephone antenna may be insignificant compared to other sources of radiation in urban environments. But at 3 inches away, things may be quite different.

Two conclusions seem warranted. First, not enough is known about the effects of low-power radio frequency (RF), and more studies are in order. Second, we should not forget that we are surrounded by many potential hazards, both natural and manmade. We not only need to determine whether the risk is real—were those brain tumors really caused by exposure to low-power RF?—but whether the risk is significant. There is considerable evidence suggesting it would be more dangerous to drive 10 miles to visit someone face to face than to call them on a handheld cell phone.

2.8 MOBILE COMMUNICATIONS STANDARDS

Nearly everyone agrees that, in the long run, technical standards benefit both vendors and end users. Technical standards may be established to enable the same mobile device to work in different locations, and to ensure maximum economy of scale for manufacturers (who are spared from developing different models for different markets). But there is considerable disagreement over when standards are needed, where they should come from, and how (if at all) they should be enforced.

At odds are two great traditions: the European tradition of mandatory standards and the American tradition of voluntary standards. This is not meant to suggest the choice is obvious. Standards have value only to the extent that people adhere to them. What is really at issue here is not the end—universal standards—but the means of producing the *best* standards.

The proponents of mandatory standards argue that, in any given market, the absence of a single standard is an obstacle to market development. Under such conditions, customers are understandably confused, and therefore hesitant to buy. Proponents of voluntary standards, on the other hand, argue that market development hinges on the quality of standards. The buying decisions of customers are a better indicator of value than the consensus of a committee of industry experts. In essence, one camp believes that *standards create markets*, while the other camp believes that *markets create standards*.

Sergio Mazza, President of the American National Standards Institute (ANSI), the official U.S. member of the International Organization for Standardization (ISO) and the International Electrotechnical Commission (IEC), rebuked the notion that government should play a decisive role in setting standards:

> Experience . . . has demonstrated that voluntary consensus has been the best method since the first standards were set for the thread sizes of machine bolts in the industrial revolution . . . Time and again, experience has proven that in the United States, the standards that have been developed through an open, nondiscriminatory, voluntary, private-sector-led, consensus process are the standards which have found the widest acceptance and the greatest utility . . . because they were generated to meet user needs.

Nevertheless, one must recognize the power and influence of various standards bodies. International standards are particularly important in the wireless realm, because both terrestrial- and satellite-based radio signals have a tendency to ignore national borders. European standards are important not only because adherence may be a prerequisite to participation in the European market, but because they often serve as the model for standards adopted in countries outside of Europe. And while compliance is often voluntary in the United States, standards tend to become increasingly important as the market develops. Below we look at some (but certainly not all) of the world's standards bodies.

The International Telecommunications Union (ITU) is a treaty organization under the United Nations that sets both technical standards and frequency allocations on a worldwide basis. The International Frequency Registration Board monitors international frequency assignments. The CCITT works under the ITU on landline standards, while the International Radio Consultative Committee (CCIR) handles radio standards. The work of the CCITT does, however, affect mobile communications; for example, one of its goals is to define how a single personal telecommunications number (PTN) could replace the multiple numbers representing a user's home, office, car phone, pager, and fax.

Every 15 or 20 years, the ITU organizes a World Administrative Radio Conference (WARC) to examine and make decisions regarding spectrum management issues. Special WARCs, focused on specific subjects, are organized more often. WARC'92 was held in February 1992 in Torremolinos, Spain and focused on future public land mobile telecommunications service (FPLMTS), mobile satellite service (MSS), DBS, and digital audio broadcasting (DAB).

In the United States, the Telecommunications Industry Association (TIA) is an association of manufacturers that sets a number of telecommunication interface standards, particularly for analog and digital cellular telephone. Also not to be ignored in this area is the Cellular Telephone Industry Association (CTIA), an association of cellular telephone service providers.

The Institute of Electrical and Electronics Engineers (IEEE) is a professional society that has played a major role in defining LAN standards. The 802.11 committee is specifically developing standards for wireless LANs.

Bellcore is a research organization jointly owned by the seven Regional Bell Operating Companies (RBOC). Bellcore seeks to develop standards that will benefit the telephone industry, and has proposed the universal digital personal communication system (UDPCS) architecture for personal communications, based (not surprisingly) on a hybrid wireless/wireline network. In Bellcore's UDPCS scheme, the PSTN provides the intelligent network features required for roaming, authentication, and billing. Another Bellcore effort, wireless access communications systems (WACS), defines fixed wireless access (FWA) to LECs as an alternative to cable-based local loops.

A number of different organizations are attempting to develop PCS standards for the United States. Some parties, such as AT&T, would like to see the U.S. PCS standards closely modeled after its digital cellular standards. Others would like to see the United States emulate Europe's global system for mobile (GSM) communications standard. The PCS Technology Advocacy Group (PTAG) includes Bell Atlantic, BellSouth, Pacific Bell, Time Warner, US West, and Canada's Stentor. The Joint Technical Committee on Wireless Access (JTC) is coordinating PCS standards efforts between the TIA, the Exchange Carriers Standards Association (ECSA), and the Personal Communications Industry Association (PCIA). But only one thing is certain: the United States is a long way from settling on PCS technical standards.

The Personal Computer Memory Card International Association (PCMCIA) has defined credit-card size plug-in *PC cards*. Though originally defined exclusively for add-on memory, the PCMCIA standard has been adapted (version 2.0) to requirements such as wireless communications and is expected to become the de facto standard. The PCMCIA was founded in June of 1989, has over 240 members worldwide, and released version 2.0 of its specification in September of 1991.

Canada's Department of Communications (DOC) is a government agency similar to the FCC in the United States, except that the DOC takes a more active role in setting cordless standards.

The situation in Europe is more complex. Until recently, the post, telephone, and telegraph (PTT) authorities were government agencies and the sole providers of telephone service and customer equipment in most European countries. Several countries have taken steps to spin off government-owned telecommunications services and introduce competition. This is not, of course, an easy process.

The Committee of European Posts and Telecommunications administrations (CEPT) is an organization of western and eastern European PTTs. Its Coordinating Committee for Radiocommunications (CR) sets frequencies for pan-European services. The Service and Facilities Committee (SF) defines all public and certain private services. CEPT defined the first- and second-generation cordless standards (CT1 and CT2) and associated pan-European spectrum allocations.

The European Telecommunications Standards Institute (ETSI) was formed in 1988. It was created to bring together a wider group than encompassed by CEPT by bringing in equipment manufacturers and non-PTT service providers. ETSI is supported by member fees, is headquartered in southern France, and has taken over much of CEPT's responsibilities in defining technical standards. For example, ETSI is defining the HIPERLAN wireless LAN; CEPT will recommend its pan-European frequency allocation.

One can point to cases that appear to support either the committee-driven approach or the market shootout approach to standards setting. What is clear is that committee-generated standards almost always leave details for "further study" or vendor discretion. And market shootouts are rarely resolved without the formation of alliances. But overall, committees seek compromise rather than innovation. Without a constant influx of bold new ideas, there is little hope of a personal communications revolution.

2.9 FCC PART 99 PCS RULES

The FCC allocated 160 MHz of spectrum in the 1.8- to 2.2- GHz region to broadband PCS in September 1993. More specifically, the FCC created a "low" band from 1,850 to 1,970 MHz and a "high" band consisting of two segments, 2,130 to 2,150 MHz and 2,180 to 2,200 MHz. A narrowband PCS (consisting of 50- and 12.5-kHz channels) was also established in the 930-MHz band for applications such as two-way paging.

Of the 160 MHz allocated for broadband PCS, 120 MHz were assigned to licensed PCS and 40 MHz to UPCS. The licensed spectrum was divided into two 30-MHz, one 20-MHz, and four 10-MHz blocks. The unlicensed spectrum was divided into two 20-MHz blocks, with 1,890 to 1,900 MHz and 1,920 to 1,930 MHz for isochronous applications (such as voice), and 1,900 to 1,920 MHz for asynchronous applications (e.g., data). But UPCS would be more accurately called *licensed unlicensed PCS*, because users are restricted (at least until all incumbent microwave users are relocated) to nonnomadic use (i.e., fixed locations) on coordinated frequencies.

Ten-year PCS licenses will be awarded, through competitive bidding, in major trading areas (MTA, each about the size of a state) and basic trading areas (BTA, each about the size of a city) as defined by Rand McNally Corporation. There were also build-out requirements (e.g., must offer service to one-third of the population within 5 years), eligibility restrictions (e.g., cellular licensees can only compete for one 10-MHz block in their existing service area), and proposed preferential treatment for small businesses, woman- or minority-owned businesses, and rural telephone companies (with one 10- and one 20-MHz block targeted for study).

This grand compromise was met with a torrent of criticism from virtually every quarter. PCS advocates complained the new service contained too many unattractive (from the viewpoint of potential investors) 10-MHz blocks and lacked the necessary contiguous spectrum to compete with cellular. In a decision handed down in June of 1994, the FCC revised the licensed PCS allocation to consist of three 30-MHz blocks and three 10-MHz blocks in the 1,850- to 1,910-MHz and 1,930- to 1,990-MHz bands. Unlicensed PCS—already in doubt due to prohibitions against nomadic use—was all but killed as the FCC "temporarily" reduced its allocation to a mere 20 MHz (Figure 2.2).

FCC Rulemaking PP Docket 93–253 stipulates that auctions must be employed when companies file mutually exclusive applications for initial licenses or construction permits, and when licensees will have paying subscribers. Excluded are commercial TV and radio broadcasting. Included are SMR, interactive video data service (IVDS), and other CMRSs. Both narrowband and broadband PCS licenses are to be auctioned off.

So what does all of this mean? Prospects for the new PCS are not good. The FCC has allowed itself to become a battleground upon which every special interest group has converged. Based on the cellular lottery experience a decade ago, a naive

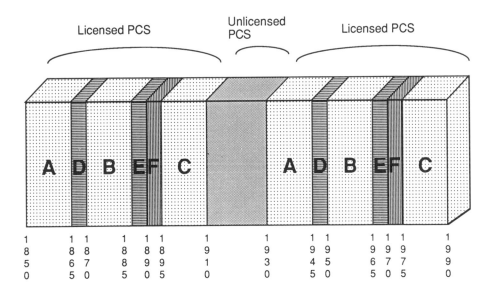

Frequency in MHz

Figure 2.2 Broadband PCS spectrum allocation (June 1994).

public believes the FCC can create "sure thing" investment opportunities by decree. Most view PCS as an opportunity to grab a part of the burgeoning cellular telephone market. And if they can get the government to give them a piece at a discount, or tie the hands of who should have been their most formidable competitors, so much the better.

PCS should have been a totally new service—wide open to innovative new technologies. It should have been defined with as few rules and restrictions as possible. The FCC should have challenged the industry to produce technological solutions for effective band sharing and interference avoidance. And the focus should have been on encouraging broadband voice, data, and video applications—applications never before seen in the radio spectrum.

Fortunately, the personal communications revolution does not depend on PCS. As we shall see, personal communications is percolating throughout the electromagnetic spectrum, and there is nothing the politicians can do to stop it.

SELECT BIBLIOGRAPHY

[1]   Brodsky, I., "Interview With Benn Kobb," *The Wireless Industry Prospectus*, Vol. 1, No. 1, 1992.

[2]   De Sola Pool, I., *Technologies of Freedom: On Free Speech in an Electronic Age*, Harvard University Press, 1983.

[3]   Gilder, G., "Auctioning the Airwaves", *Forbes ASAP*, April 11, 1994.

[4]   Shooshan, H. M., III, *Disconnecting Bell*, Pergamon Press, 1984.

[5]   Press announcement from ANSI, Rockville, MD, June 15, 1994.

[6]   *U.S. Spectrum Management Policy: Agenda for the Future*, National Telecommunications and Information Administration, Washington, DC, 1991.

[7]   *Telecommunications in the Age of Information*, National Telecommunications and Information Administration, Washington, DC, 1991.

[8]   *Telecommunications Reports Wireless News*, BRP Publications, Washington, DC 1993, 1994 (biweekly).

CHAPTER 3

THE MYSTERIES OF
SPREAD SPECTRUM

3.1 THE WIRELESS STONE AGE

For decades, radio engineers were taught that the radio spectrum is a scarce resource. Designing mobile communication systems that consume as little bandwidth as possible became standard engineering practice. Even the FCC participated in the bandwidth conservation movement, splitting 30-kHz channels into 15-kHz channels and establishing new services composed of even narrower slices. For example, the 220-MHz band in the United States—recently reassigned from amateur to commercial mobile radio—has been divided into tiny 5-kHz parcels.

While mobile users are admonished to get by with less, the bandwidth offered to landline users has grown by leaps and bounds. Despite critics' assertions that they could not possibly know what to do with it, tethered users continue to invent bandwidth-guzzling applications. Today, landline users contemplate gigabit-per-second networks, while mobile users are told that 5-kHz-wide channels represent the state of the art. This enormous—and growing—disparity raises serious questions about the long-term prospects for mobile communications.

The view that wireless users must learn to ration bandwidth, however, stems from a fundamental pessimism regarding technology. Pessimists tend to believe that the good life is coming to an end; we must finally learn to live within our means. Their solution to the spectrum shortage is a bandwidth austerity program. But just as many were unwilling to accept "brownouts" as the solution to the energy crisis, optimists believe we can conquer the spectrum shortage. Where pessimists see only insurmountable problems, optimists see opportunities.

Great inventions often turn conventional wisdom on its head. For example, during the early days of solid-state electronics, designers used as few (discrete) transistors as possible. When ICs were first conceived, it proved extremely difficult to match the performance of a discrete transistor when the component was scaled down to fit inside an IC. The solution was to employ designs that (by existing standards) wasted transistors. Now, electronic devices containing ICs with millions of transistors are common. Similarly, could the solution to the spectrum shortage be a technology that (by existing standards) wastes bandwidth?

A technology called *spread spectrum* does precisely that. Once shrouded in military secrecy, spread spectrum radio increases capacity not by confining each user to the narrowest possible frequency slice, but by bestowing each user with significantly more bandwidth than he or she appears to require. Although the concept has been around for decades—it was originally conceived to thwart enemy jammers—only within the past decade have components emerged that are propelling spread spectrum into widespread use.

3.2 WIRELESS MAGIC

Three types of interference plague radio. Nearby users on the same or an adjacent channel may cause interference. Radio signals taking different paths from the transmitter to the receiver may interfere with each other. And jammers produce interference with the intent of disrupting others' communications. In its various guises, interference limits the security, reliability, and capacity of radio communications.

In contrast, tethered users encounter relatively little interference. By confining signals within strands of copper or glass, the same bandwidth may be reused over and over. The aggregate capacity of metallic and fiber-optic cables, even when run side by side, is enormous. But because wireless signals are propagated through free space, it is as if all of the users within a given area were forced to share a single cable.

However, if we could somehow suppress interference, wireless would begin to look more like wired communications. Spectrum capacity would be greatly increased, since we would be able to reuse bandwidth aggressively. Transmitters would become as abundant as receivers—everyone would own at least one; most people would own several. Because spread spectrum has a unique ability to suppress interference, it promises much greater spectrum capacity.

Of course, other developments have also helped increase spectrum capacity. Trunked radio enables a group of users to share a smaller number of channels on a contention basis. Cellular radio allows users to reuse the same channels over a geographic area. Time-division multiplexing allows a group of users to share the same channel by taking turns. But only spread spectrum permits us to distribute traffic smoothly over an entire radio band.

Most of the world is still betting on more conventional digital radio techniques: frequency-division multiple access (FDMA) and time-division multiple access (TDMA). Experts believe that FDMA and TDMA are less complex and are therefore best suited to be the first-generation digital radio solution. Many concede that spread spectrum may ultimately yield greater capacity, but suggest it be considered as a second- or third-generation solution.

Indeed, complexity is spread spectrum's main drawback. Complexity usually translates into higher cost and poorer reliability. But in an era of very-large-scale integration (VLSI), cost and reliability problems may vanish quickly. Once a solution is embedded in semiconductor wafers, it becomes a commodity. Spread spectrum chips will serve as building blocks to a wide variety of products. Besides, implementation of TDMA—the most popular first-generation digital radio solution—also turns out to be more complex than had been anticipated.

Skeptics should consider the history of high-speed dial-up modems. The CCITT V.32 standard achieves full-duplex, 9,600-bps communications over a 3,300-Hz voice circuit by employing a highly complex technology known as *echo canceling*. When the first V.32 modems hit the street, many questioned whether they would ever become affordable. Indeed, some wondered if they would ever work as promised. The first V.32 modems appeared around 1987, cost upwards of $4,000 each, and generally did not work with units made by other vendors despite the existence of a standard. But by 1993—just 6 years later—V.32 modems not only worked, but were widely available for as little as $150, often incorporating a number of powerful enhancements. Siliconization yielded a more than twentyfold reduction in price—not to mention a considerable improvement in performance.

Perhaps spread spectrum is not radio's "magic bullet." Then again, perhaps it is.

3.3 WHAT IS SPREAD SPECTRUM?

As its name implies, spread spectrum radio works by spreading signals over a wide frequency band. More precisely, spread spectrum signals must be spread (1) over a much wider frequency band than would ordinarily be required by their information content, and (2) by combining a pseudorandom signal with the original information-bearing signal.

When a conventional narrowband signal is combined with a wideband pseudorandom signal, the end product resembles white noise. Therefore, spread spectrum signals look like background noise to conventional (narrowband) receivers. In fact, spread spectrum is sometimes referred to as *noise modulation*. Only a receiver possessing the correct pseudorandom spreading code can recover the original user information. It does so through a process known as *correlation*, in which the received signal is lined up with and multiplied by the proper pseudorandom code.

Here is where things get really interesting. A moment ago, we said that when you combine a narrowband signal with a wideband pseudorandom signal the resulting signal looks like white noise. But this is precisely what the receiver's correlator does to any narrowband signals it encounters. In other words, the spread spectrum receiver converts a noiselike spread spectrum signal back into a narrow-band signal, and at the same time converts any narrowband signals received into white noise.

Claude Shannon, the father of modern communication theory, demonstrated that at a fixed rate of information flow, one can always trade signal-to-noise ratio (SNR) for bandwidth. In other words, it is possible to operate at a lower SNR if a channel's bandwidth is increased. Shannon was most interested in how special cod-ing techniques—which expand a signal's bandwidth by adding redundant informa-tion—could be exploited when operating at low SNRs.

Although Shannon was not investigating spread spectrum, notice that pseu-dorandom spreading codes achieve a similar effect. Spread spectrum signals operate at a much lower power density than conventional narrowband signals. In fact, spread spectrum signals have a mind-boggling ability to operate close to—or even below—the noise floor. For this reason, spread spectrum signals may be difficult to detect by those not tipped off to their presence.

There are several different ways to spread a signal. One method is to simply widen the signal out over a continuous range of frequencies. This technique, implicit in our discussion so far, is called *direct sequence spread spectrum* (DS-SS) and is illustrated in Figure 3.1.

We can also spread a signal by causing it to hop rapidly from one frequency to another. In this case, the narrowband user information signal remains intact but hops from channel to channel according to a predetermined pseudorandom se-quence. This technique is called *frequency-hopping spread spectrum* (FH-SS) and is illustrated in Figure 3.2.

There are other techniques such as chirp spread spectrum, in which the signal is swept across a frequency band. But DS-SS, FH-SS, and hybrid DS/FH-SS are by far the most popular approaches.

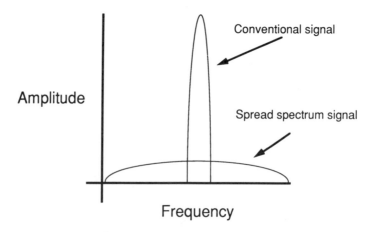

Figure 3.1 Direct sequence spread spectrum.

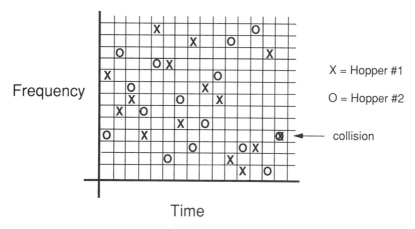

Figure 3.2 Frequency-hopping spread spectrum (asynchronous).

3.4 A HOLLYWOOD STARLET OR THE U.S. AIR FORCE?

The invention of spread spectrum cannot be traced to any one person or event. Spread spectrum–like radio systems were conceived as early as the 1920s. An often told story concerns the movie actress Hedy Lamarr. Hedwig Eva Maria Kiesler (her real name) left Austria in 1938 for the United States, where she became close friends with symphony composer George Antheil. Although she frequently played the role of a vacuous blonde, Lamarr was an intelligent woman who fervently desired an Allied victory in World War II. Having learned about warfare from her former husband, an Austrian munitions merchant, Lamarr, together with Antheil, conceived a radio-controlled system for guiding torpedoes that would be "veritably impossible for enemy vessels to jam . . ."

In the Lamarr/Antheil system, the transmitter on the attacking ship or submarine hopped from frequency to frequency according to a prearranged, randomized sequence. The torpedo was equipped with a receiver synchronized to the same hopping pattern. Drawing inspiration from the paper rolls used in player pianos, the couple obtained one of the first patents for what we now call FH-SS.

In reality, the Lamarr/Antheil patent is just a footnote in the history of spread spectrum. The idea as conceived was never implemented, nor did the patent contribute towards the future development of spread spectrum radios. It became important when it was rediscovered years later simply because it provided a colorful story.

But the patent does bring to light a problem that plagued all early spread spectrum systems. Prior to the development of ICs, it was difficult to generate pseudorandom codes of sufficient length. As in the Lamarr/Antheil approach, most early systems employed some sort of electromechanical device to generate the pseudorandom code. For example, Federal Telecommunication Laboratories (FTL) used *noise wheels*, in which light from a slit lamp was modulated by passing it through a circular piece of film rotating in front of a photoelectric cell. The receiver either had to have an identical and synchronized noise generator, or the noise signal had to be transmitted separately (an approach that could jeopardize confidentiality).

The first undisputed spread spectrum radios appeared in the late 1950s—made possible with the introduction of transistors. The coining of the term *spread spectrum* is credited to Madison Nicholson and John Raney of Sylvania Buffalo. The duo headed up a military project dubbed "hush-up" which led to the construction of the first airborne spread spectrum radio system, the ARC-50.

The prototype ARC-50 was flight-tested at Wright Patterson Air Force Base. The ground terminal was located about 100 yards from the base's control tower and operated over frequencies overlapping those of the tower. Although the ARC-50 communicated with aircraft up to 100 miles away, tower personnel were never aware the spread spectrum transmissions were taking place.

The ARC-50 was eventually manufactured by Magnavox. It consisted of two boxes, one a UHF radio and the other a spread spectrum modem, which together contained roughly 600 transistors. Unfortunately, a single ARC-50 consumed the chassis slots normally reserved for two vacuum tube radios. In those days, the standard vacuum tube Air Force radios had a mean time before failure (MTBF) of about a half hour. If a pilot wanted a reasonable chance of receiving landing instructions upon returning to base, he had to take two radios. Even though the transistorized ARC-50 had an MTBF of about 400 hours, pilots resisted its use, because they still felt they needed two radios. Most of the 1,000 ARC-50s manufactured in 1961 ended up in a warehouse.

The first commercial applications of spread spectrum did not emerge until the late 1970s. These included air-to-air collision avoidance, very-small-aperture terminal (VSAT) satellite, and commercial global positioning system (GPS) receivers. The first mainstream commercial application was a low-cost satellite-based data network developed by Equatorial Communications Company in 1979. At that time, conventional technology dictated 15-ft-diameter dish antennas, because satellites were spaced just 4 degrees apart in the sky. (Today, they are even more tightly packed: one every 2 degrees). But because of its ability to attenuate interference, spread spectrum could be used with 2-ft-diameter antennas—a development that some believe launched the VSAT industry. This use of spread spectrum, by the way, was accomplished under FCC rules which neither permitted nor prohibited its use.

3.5 DIRECT SEQUENCE SPREAD SPECTRUM

Direct sequence is probably the most popular type of spread spectrum. It is also easy to understand. Imagine a user data stream running at 8,000 bps. Now imagine we combine this data stream with a pseudorandom spreading code running at 800,000 bps. Each bit of original end user data is now represented in the composite data stream by 100 bits (but now called *chips*). The number of chips per bit, or the degree to which the end user data are spread out, is known as the *processing gain*.

A receiver that knows the pseudorandom spreading code can work backwards to recover each original end user bit (correlation). If at any moment an end user 1 is represented by a particular sequence of chips, then an end user 0 would be represented by a complementary sequence in which each 1 chip has been replaced by a 0 chip, and vice versa. Notice that the receiver does not need to be perfectly accurate. If it correctly recognizes just 51 of the 100 chips representing a given user bit, it will still come up with the right answer. In the world of (end user) binary data, each bit is extremely important; in the world of spread spectrum binary chips, the majority rules.

DS-SS has several advantages. A direct sequence signal may be hidden from any receiver—including a spread spectrum receiver that does not possess the correct spreading code. Multiple DS-SS signals based on orthogonal (i.e., sufficiently different) spreading codes may share the same wideband channel at the same time. And DS-SS signals are relatively immune to multipath fading. The more a DS-SS signal is spread out (the higher the processing gain), the more pronounced each of these advantages becomes.

Amazingly, DS-SS systems can communicate at signal levels below the ambient noise level. During the correlation process, the desired signal is improved while ambient noise remains steady. For example, the GPS routinely works at 31 dB below the noise level. When a signal that has been spread to 10 MHz is correlated back down to 100 Hz, a 50-dB improvement (denoted by 50 dB of processing gain) is obtained.

But DS-SS also has its disadvantages. In a given bandwidth, the end user data rate is more constrained than would be true with conventional radio techniques. For example, the maximum end user information bit rate achievable using DS-SS in accordance with FCC Rules in the 902- to 928-MHz ISM band is approximately 2 million bps, even though the available bandwidth is 26 MHz. Another disadvantage of DS-SS, especially when compared to FH-SS, is that the presence of a sufficiently strong interferer anywhere in the (wideband) channel can disrupt communication. The maximum tolerable difference in signal strength between an interfering narrow-band signal and the DS-SS signal is called the *jamming margin*.

3.6 CODE-DIVISION MULTIPLE ACCESS

One of the most fascinating properties of DS-SS is its ability to support multiple simultaneous users on the same wideband channel—what is known as code-division multiple access (CDMA). Each CDMA receiver recovers only the end user information it is looking for; the other DS-SS signals continue to look like white noise. Unlike conventional multiple-access schemes (e.g., frequency division or time division), CDMA makes no attempt to segregate users into distinct channels or time slots.

This is where the fact that a DS-SS receiver does not need to correctly recognize every single chip comes in handy. A single base station transceiver can communicate simultaneously with multiple mobile users. Outbound transmissions can be combined into a composite chip stream in which each mobile receiver will still see a majority of chips representing the correct bits, even though now it can be ensured that some of the transmit chips will appear to be incorrect. The same thing happens, in essence, at the base station receiver, which sees a composite chip stream from all active mobile transmitters, yet can still recover the correct bits for each based on the "majority wins" rule. (It may help to imagine a simple CDMA system with just two simultaneous conversations and employing seven chips per bit. Each of the receivers simply needs a majority of chips—i.e., four or more—to match its correct chip sequence. As long as the two codes are orthogonal, this should always be possible.)

Another advantage of CDMA is that, unlike FDMA or TDMA, there is no "hard" capacity limit. In an FDMA system, for example, once all of the frequency channels are filled, no additional users can be served without causing severe interference. In a CDMA system, however, each additional user creates an incremental increase in white noise for all users of the wideband channel. CDMA proponents boast that this "graceful degradation" makes the system self-regulating: noting the high noise level, some users will decide to defer their communications until the system is less busy.

By now, CDMA must sound like magic. But it has its Achilles' heel: the *near-far problem*. In order for CDMA to work properly, each of the signals must be received at the same signal strength. If one transmitter is significantly closer to the receiver than all others, its signal may swamp the receiver's front-end and prevent the reception of the other signals. CDMA can only be used in a mobile communications network by employing reverse power control. That is, by dynamically controlling the transmit power level of every mobile station, a CDMA base station can make certain that all of the signals are received at equal strength—regardless of how near some mobile stations may be. Many experts believe the complexity and added cost of dynamic power control is CDMA's fatal flaw (Figure 3.3).

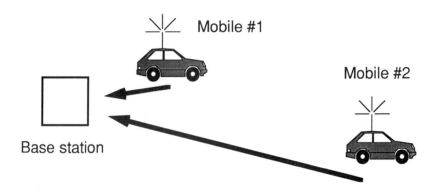

Figure 3.3 The near-far problem.

3.7 FREQUENCY HOPPING

Instead of continuously spreading the signal over a wideband channel, a frequency hopper is a conventional signal (i.e., one that occupies only the bandwidth required by its information content) that jumps from frequency to frequency. But FH-SS has an important advantage over DS-SS: it is more robust in the presence of powerful narrowband interference. In fact, if the frequencies of narrowband interferers are known in advance, an FH-SS system can simply select a hopping sequence that does not include those frequencies. In other words, while DS-SS systems attenuate interference, FH-SS systems avoid it.

There are several different flavors of FH-SS. Fast-hopping FH-SS systems use multiple frequency hops for each bit of end user data. Like DS-SS, the fast-FH-SS receiver does not need to get every hop (chip) right. A fast-FH-SS receiver can even tolerate a number of unexpected narrowband interferers, as long as it still gets the majority of chips correct.

Fast hoppers are particularly effective at thwarting jammers. One ingenious FH-SS system signals a 0 chip or 1 chip through the presence or absence of a radio carrier. The receiver dutifully travels from hop to hop (according to a prearranged pattern) to see if the carrier is present. Imagine the fits this must cause enemy jammers! Even if they have the ability to instantaneously find and jam the carrier, it will be to no avail. If a 1 chip is signaled by the absence of carrier, there is nothing for the jammers to find. If they successfully locate and jam a frequency on which a 0 chip is being signaled, their jamming signal will only reinforce the carrier.

Under specific conditions, it is physically impossible to jam a fast-hopping FH-SS system. This occurs when the (friendly) receiver is closer to the (friendly) transmitter than the enemy transmitter (distance C is less than B), and the duration of each hop is less than the time it takes for the signal to propagate from the (friendly) transmitter to the enemy (propagation time A) (Figure 3.4). Even if the

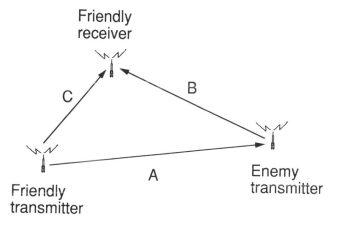

Figure 3.4 A jam-proof FH-SS scenario.

jammers find and jam each hop instantaneously, they are too late. By the time the enemy hears the signal, the (friendly) transmitter and (friendly) receiver have already hopped to the next frequency.

In contrast, a slow hopper sends more than one bit of end user information per hop. The advantage of this type of system is that it is less complex and therefore less expensive to build. While the military prefers fast FH-SS, most commercial developers prefer slow FH-SS.

3.8 FREQUENCY-HOPPING MULTIPLE ACCESS

As with DS-SS, multiple FH-SS users can share the same wideband channel at the same time. There are two types of such frequency-hopping multiple access (FHMA) systems. In a synchronous FH-SS system, the hopping sequences of multiple users are synchronized so that no two users ever transmit on the same frequency at the same time. The advantage of this system is that one can guarantee a minimum data rate for each user. The disadvantage is that only a limited number of synchronous hoppers can coexist over the same wideband channel.

Critics argue that a synchronous FH-SS system is really nothing more than a collection of narrowband users playing musical chairs. In fact, if the number of users equals the number of available hopping channels, and each transmitter operates with a 100% duty cycle, there is no advantage (other than privacy) over a conventional system in which each user occupies a single channel all of the time.

In an asynchronous FHMA system, the hopping sequences are not synchronized and there is always a chance that two or more users will collide, resulting in bit errors. Critics point out that if there are two pairs of users on a 1,000-channel system, the chances of a collision are 1 in 1,000. This yields a BER of 1×10^{-3}—an unacceptably high error rate for most data applications. And with 100 users, the BER would skyrocket to 1 in every 10 bits.

Proponents point out that these nightmare scenarios have little to do with real applications. In most packet data applications, transmitter duty cycle is well below 100%. Though there may be 100 users on a system, not all transmit at the same time. If nothing else, keeping the users moving (frequencywise) helps to minimize the impact of narrowband interferers (as long as they are not too numerous) and multipath fading.

3.9 APPLICATIONS OF SPREAD SPECTRUM

There are three major groups of applications for spread spectrum radio: military, business, and consumer. We have already discussed why the military is interested in spread spectrum. Most business and consumer applications are made possible by government regulations that permit end users to purchase and use spread spectrum devices without obtaining operator's licenses.

In the United States, unlicensed use of spread spectrum devices is covered under Part 15 of the FCC Rules. In the late 1970s, the FCC was jolted by a scathing report from the General Accounting Office (GAO) criticizing it for not planning ahead, especially with regard to new technologies. About this time, Mike Marcus came to the FCC from the Institute for Defense Analyses. The FCC was anxious to prove it could respond to new technologies, and Marcus saw the potential of spread spectrum.

Hewlett-Packard (HP) was perhaps first to petition the FCC to allow unlicensed use of spread spectrum. HP was looking for a portable network—one that was easy to pack up and move—for training seminars it conducted around the country. The initial NPRM was adopted in 1981. Predictably, there was fierce opposition from incumbent land mobile users. The proposal outlined three permitted uses of spread spectrum radio: (1) surreptitious tracking and monitoring for law enforcement, (2) experimentation in the amateur radio service, and (3) operation in the ISM bands under FCC Rules Part 15 at transmit power levels up to 1W. These Rules were finally adopted in 1985 and revised in 1989.

Spread spectrum devices operating under Part 15 Rules in the ISM bands include wireless LANs, wireless office telephone systems, and (most recently) enhanced range/privacy residential cordless telephones. The ISM bands were created to accommodate devices that use radio frequency energy for purposes other than communications, such as microwave ovens or jewelry cleaners. ISM devices and government users have top priority in the ISM bands, followed by licensed automatic vehicle monitoring (AVM) systems. Part 15 devices and amateur radio operators must not interfere with, and must tolerate interference from, these users. In other words, spread spectrum wireless LANs and wireless PBX adjuncts operating in the ISM bands can be forced to shut down if they interfere with priority users—a fact little known among users of these devices. Fortunately, such conflicts have been extremely rare, and in most instances are resolvable.

The Part 15 Rules permit direct sequence, frequency-hopping, and hybrid spread spectrum systems to operate at up to 1W of transmit power—much more power than non-spread spectrum devices are permitted. For example, conventional residential cordless telephones are allowed to transmit roughly 0.75 mW. The FCC agreed to allow spread spectrum devices to transmit at higher power levels because it was convinced of spread spectrum technology's lower probability of causing interference.

The first spread spectrum product certified for use under Part 15 was in 1987—a wireless LAN developed by Telesystems SLW of Canada. Shortly after, the number of firms developing Part 15 spread spectrum products skyrocketed. One reason was the emergence of low-cost, high-density ICs such as field-programmable gate arrays containing 2,000 to 4,000 gates. Today, a single application-specific integrated circuit (ASIC) can handle the functions of spread spectrum modulator, demodulator, code generator, and correlator. In fact, it is now possible to design a complete spread spectrum radio using just two chips—the Stanford Telecom STEL-

2000 10-megachip/second spread spectrum chip and the GEC-Marconi 2.4-GHz RF chip—and a handful of small parts.

Another reason for Part 15 spread spectrum' growth is that vendors want to take advantage of the ability to market unlicensed products that can transmit at power levels of up to 1W. Indeed, many products use the minimum required processing gain. Critics warn that as the installed base of such products grows, so too will interference problems. But because most Part 15 spread spectrum products are portable devices, many operate at less than 1W to conserve battery power. There are no signs of an imminent crisis caused by the proliferation of unlicensed spread spectrum devices.

A potentially major application for spread spectrum can be found in cellular and microcellular networks (PCS). The CDMA version scored a major coup in the U.S. cellular telephone industry. Although North America had chosen TDMA for the official standard for digital cellular networks, Qualcomm and its supporters have managed to position CDMA as an officially recognized alternative standard. Several manufacturers have committed to manufacture CDMA equipment, and cellular carriers such as US West and AirTouch Cellular (previously Pactel Cellular) have decided to bypass TDMA and deploy CDMA instead.

There are two reasons for the cellular industry's interest in CDMA. First, CDMA offers greater capacity than TDMA in cellular networks. Although CDMA is less efficient in an individual cell, it allows frequencies to be reused in adjacent cells. (TDMA does not normally permit the reuse of frequencies in adjacent cells.) We will return to this subject in Chapter 4.

But perhaps a more important reason why cellular carriers are interested in CDMA is political rather than technical. The new Personal Communications Service Rules have been designed to restrict a cellular carrier's participation in its own backyard—areas in which it already provides cellular service. Supporters of such restrictions believe that cellular carriers, by virtue of their existing network infrastructure and customer base, possess an unfair advantage. (An alternative interpretation is that some people would like the government to clear a path for them to grab a share of the mobile telephone market.)

But there is no reason why cellular carriers could not provide PCS-like services using their existing spectrum. Some carriers see CDMA as a vehicle for doing precisely that, because it offers greater capacity than TDMA and simplifies network planning. (There is no need to coordinate CDMA channel assignments between cells). In any event, it would not be the first time that technology enabled businesses to circumvent barriers created by regulators.

For a long time, it was a foregone conclusion that spread spectrum would be the dominant PCS technology in the United States. The first experimental PCS license was awarded to PCN America, a subsidiary of Millicom, which had teamed up with spread spectrum pioneer SCS Mobilcom (later acquired by InterDigital Communications Corporation). The consensus seemed to be that spread spectrum would enable PCS and incumbent fixed microwave users to coexist; FCC-sponsored field tests, however, were indeterminate since both sides claimed victory. (Of course, most incumbent users were less interested in sharing and more interested in getting paid to relocate elsewhere in the spectrum.)

While spread spectrum may still be the superior technology, the FCC's broadband PCS auctions may have the unintended effect of squeezing out the entrepreneurs who were most receptive to innovative technologies. It will be difficult for medium-size firms to outbid the interexchange carriers and RBOCs on the one hand, and those receiving preferential treatment in the way of discounts on the other. (These include rural telephone companies, women-owned, minority-owned, and small businesses.) Furthermore, the astounding success of the narrowband PCS auctions—winning bids totaled more than $600 million—can only encourage the FCC to try to delay migration to technologies that might render exclusive licenses obsolete.

3.10 FOUR SPREAD SPECTRUM PIONEERS

Perhaps the reason spread spectrum is certain to succeed is, in fact, that it has captured the imaginations of a small army of entrepreneurs. Let's take a look at four spread spectrum innovators.

Omnipoint Corporation is a Colorado Springs–based firm that designs and manufactures spread spectrum radios. The firm's chief scientist, Robert C. Dixon, began working with spread spectrum in 1959 (with Magnavox on the ARC-50) and is considered one of the field's leading experts. Dixon's practical book, *Spread Spectrum Systems*, has played a major role in popularizing the technology.

Dixon joined Omnipoint to develop spread spectrum–based products for PCS. Although spectrum sharing tests conducted by Millicom were deemed inconclusive, Dixon declares, "We can share [frequency bands] with certain constraints." For example, he points outs that a 1W spread spectrum transmitter placed directly in front of a microwave antenna would still cause interference.

Omnipoint has developed a hybrid TDMA/DS-SS solution. Unlike the approach employed by Millicom during its tests, Omnipoint uses a relatively narrow DS-SS signal and first attempts to find a clear channel. Dixon's attitude is that spread spectrum is not a panacea, but will enable more users to be squeezed into what appears to be a fully used band. Omnipoint's system—which has been used by Ameritech, Bell Atlantic, and others in PCS and related experiments—currently supports 32 simultaneous users running 8 kbps each on a TDMA basis. In other words, spread spectrum is used primarily to minimize interference between Omnipoint's system and any nearby conventional systems (such as point-to-point microwave links).

Another firm, Metricom (Los Gatos, California), has developed a metropolitan area mesh network that uses asynchronous FH-SS. Designed for wireless data applications, the network (known as Ricochet) is a relatively high-speed, low-cost wireless data solution initially targeting campus environments. While nationwide mobile data networks such as those operated by ARDIS and RAM Mobile Data typically charge users for each packet of data, and operate at speeds well below what is commonly available at the desktop, Ricochet has been launched in Northern California at flat monthly rates for service at speeds of up to 56 kbps. The firm is publicly traded and has attracted investors such as Microsoft cofounder Paul Allen.

Metricom's Director of Engineering, Mike Pettus, asserts that "The key advantage to spread spectrum is that it leverages the power of the microprocessor to control a frequency-agile radio." Indeed, each radio in a Ricochet network uses the same hopping pattern, but with random synchronization. When a radio is installed, it is programmed with its precise geographical coordinates. When powered up, the radio automatically learns the identities and locations of each of its neighbors. The best route can then be automatically selected; links or nodes that are performing unsatisfactorily can be automatically avoided. Metricom's FH-SS system runs 76,920 bps over 163 channels spaced 160 kHz apart.

Ricochet nodes are self-routing packet radios that cost approximately $700 each and are small enough to be mounted on utility poles or even lamp posts. Mobile units have been developed for use with a wide variety of portable computers (Figures 3.5 and 3.6). Packets hop from node to node until they reach the desired destination (usually the nearest node with a phone line attached). Coverage or capacity (or both) may be expanded by merely installing additional nodes.

Qualcomm (based in San Diego) is probably the best known proponent of CDMA. The firm's OmniTRACS mobile satellite service uses a CDMA mobile uplink. Although OmniTRACS is financially successful, Qualcomm has attracted much more attention for its CDMA digital cellular technology, which promises 10 to 15 times the capacity of current (analog) cellular systems and approximately twice the capacity of the North American TDMA standard (IS-54).

Qualcomm's Andrew Viterbi is one of the acknowledged gurus of spread spectrum, with over 35 years of experience. Viterbi believes that DS-SS will dominate cellular telephone, PCS, and wireless local loop applications.

Figure 3.5 Metricom's FH-SS portable radio modem.

Figure 3.6 Metricom's FH-SS node transceiver.

Qualcomm's cellular solution is a relatively narrowband version of CDMA (we will call it N-CDMA), which operates over 1.25-MHz channels in the 800-MHz cellular telephone band. Each 1.25-MHz channel supports 25–40 simultaneous users per cell. But N-CDMA requires the use of reverse power control to prevent the near-far problem. Once solved, however, the power control problem becomes a virtue: mobile devices operate consistently at minimum transmit power levels, reducing power consumption in battery-powered subscriber units.

After InterDigital Communications Corporation acquired Donald Schilling's SCS Mobilcom, the firm proposed a broadband version of CDMA (we will call it B-CDMA), which would operate on top of existing analog cellular users. Unlike Qualcomm's N-CDMA, which requires clearing out all analog cellular users over a 1.25-MHz stretch, InterDigital would use a 48-MHz-wide CDMA channel and dynamically adjustable notch filters on both the receivers and transmitters (to "notch out" the analog channels).

We will take a closer look at the three competing digital cellular technologies—TDMA, N-CDMA, and B-CDMA—in the next chapter.

3.11 THE FUTURE OF SPREAD SPECTRUM

History is filled with examples of successful technologies that critics said would not work, would cost too much, or nobody would want. It is always easier to criticize a new technology than to champion it. Like successful technologies before it, spread spectrum has compelling benefits that will continue to attract persistent entrepreneurs.

The key to personal communications is abundant spectrum capacity. Spread spectrum, as Metricom's Pettus points out, permits the traffic load to be distributed more evenly across the radio spectrum. Given enough capacity, individuals will be able to access the radio spectrum for purposes once unthinkable.

But spread spectrum is a complex technology that requires further development to reach its full potential. There are no shortcuts to success, but once spread spectrum is firmly embedded in silicon, products will experience a rapid decline in size, power consumption, and—most important—cost.

The United States has been criticized for not proactively developing advanced technology standards. The real difference between the United States and Europe/Japan, however, is not that the United States lacks aggressiveness in developing advanced technologies—quite the contrary—but that in the United States adherence to most standards is voluntary. And it is for this reason that spread spectrum has emerged, albeit late, as a potentially superior technology for digital cellular and PCS.

SELECT BIBLIOGRAPHY

 [1] Calhoun, G., *Digital Cellular Radio*, Norwood, MA: Artech House, 1988.
 [2] Dixon, R. C., Omnipoint Corporation, interview on August 9, 1993.
 [3] Dixon, R. C., *Spread Spectrum Systems*, 2nd edition, New York, John Wiley & Sons, 1984.
 [4] Gilder, G., "The New Rule of Wireless," *Forbes ASAP*, March 29, 1993.
 [5] Holtzman, J., Rutgers University WINLAB, interview on August 10, 1993.
 [6] Marcus, M., Federal Communications Commission, interview on August 12, 1993.
 [7] Pettus, M., Metricom, Inc., interview on August 10, 1993.
 [8] Schilling, D., InterDigital Communications Corporation, interview on August 11, 1993.
 [9] Scholtz, R. A., "The Origins of Spread Spectrum Communications," *IEEE Transactions on Communications*, May 1982.
[10] Shannon, C. E., "Recent Developments in Communication Theory," *Electronics*, April 1950.
[11] Viterbi, A. J., "Spread Spectrum Communications—Myths and Realities," *IEEE Communications Magazine*, May 1979.
[12] Viterbi, A., Qualcomm, Inc., interview on August 13, 1993.

CHAPTER 4
▼▼▼

DIGITAL RADIO NETWORKS

4.1 FROM SHORTAGE TO GLUT

Unlike landline networks, the principal design goal of wireless networks is almost always to maximize capacity. Cellular radio, which permits frequencies to be reused throughout a city, has rapidly emerged as the most popular wireless network architecture. When it was first conceived, many believed cellular could deliver as much capacity as needed. A 50% reduction in cell radius should yield a 400% increase in network capacity. At least, that is how it works on paper.

But the perpetually shrinking cell is not, by itself, the answer to the spectrum shortage. The smaller the cells become, the more difficult it becomes to manage cell handoffs, the harder it becomes to control coverage (lower power transmitters must be used), and the more real estate, phone lines, and hardware must be acquired. While microcells make sense in areas with high user densities, another tool is needed—one that will reduce costs as well as increase capacity.

Fortunately, there is such a tool. Digital transmission leverages the power of the microprocessor to make more efficient use of available spectrum. This can be accomplished either in each cell or across the network as a whole. Digital transmission can be used within a cell to multiplex more users onto the available bandwidth. And digital techniques can be employed to control interference between cells—by avoiding, minimizing, or counteracting it—in turn permitting more aggressive frequency reuse.

There are, however, more advanced analog techniques for achieving the same goals. Narrower channels—limited perhaps to a minimum of 5 kHz—can be employed to squeeze more users onto a radio band. Techniques such as the *Zone Selector* developed by Dr. William Lee of AirTouch Cellular, which brings the transmit power of the base station closer to the mobile unit, can be used to reduce interference. By dividing each cell into overlapping circular zones, and piping radio signals over coax, fiber-optic cable, or microwave radio to antennas located within each zone, the Zone Selector permits the base and mobile stations to transmit at low power levels. This, in turn, reduces cochannel interference between cells reusing the same frequencies.

But most experts agree that the long-term solution is digital radio. Digital receivers regenerate each bit and are therefore able to filter out moderate amounts of noise and interference. Digital transmitters use techniques such as time-division multiplexing to permit three or more users to share a channel previously used by just one. And by permitting aggressive reuse of frequencies, digital radio networks promise up to a tenfold increase in capacity over comparable analog networks. (The first digital cellular networks, however, are delivering only about two to three times the capacity of analog networks.)

Digital radio has arrived. But the transition to digital networks is as much a political battle as a technology horse race. In the United States, cellular telephone operators must upgrade their networks without disenfranchising existing analog users. SMR operators have received permission to convert to a digital cellular architecture—provoking fear among cellular telephone operators who correctly see enhanced SMR (ESMR) carriers repositioning themselves as mobile telephone players. The new PCS will be constructed from the ground up out of digital radio technology. Who should be eligible to build and operate PCS networks and whether mandatory technology standards should be adopted have become subjects of intense public debate.

One thing is clear: if all of the digital upgrades take place and all of the planned digital networks are built, the United States (as well as other countries) will enjoy a huge surge in mobile communications capacity. In fact, the spectrum shortage in the United States could be transformed—almost over night—into a spectrum glut. For example, even a modest threefold increase in capacity would permit the U.S. cellular telephone industry to serve more than 60 million users.

At best, a spectrum glut would be only temporary. A significant capacity increase will not only lead to more users, but the availability of more bandwidth per user. There is no reason to believe wireless users will remain content with exclusively narrowband services, and vendors are beginning to see this, too. For example, McCaw Cellular Communications' vice chairman Wayne Perry predicts his company will use the nascent PCS to offer broadband data services.

4.2 DIGITAL DOLLARS

Network operators are going to adopt digital cellular technology only if it can help them make money. Fearing an imminent capacity crisis, the North American cellular telephone industry accelerated the development and adoption of a digital cellular standard to replace the advanced mobile phone system (AMPS) analog standard. In so doing, they knowingly deferred non-capacity-related enhancements such as data services. But it was gradually discovered that a large percentage of new subscribers were either consumers (who primarily used their cell phones during evenings and weekends) or more cost-conscious business users (who used their cell phones more sparingly). So far, few U.S. carriers are rushing their conversion to digital.

In Europe, the motivation is somewhat different. The GSM digital cellular standard is replacing a hodgepodge of incompatible national cellular networks. Cellular subscribers in the United States have always been able to use the same phone anywhere in the country—even if incoming calls still do not automatically follow most users as they roam. For the pan-European business traveler, however, GSM is the first practical cellular telephone service. And because GSM is being implemented in a new frequency band, it represents much-needed competition.

But a new motivation for upgrading to digital cellular looms in the United States. The PCS threatens to steal much of cellular telephone's growth, if not existing market share. PCS infrastructure, based exclusively on digital cellular and microcellular architectures, should offer greater capacity, more features, and better operating efficiency than analog cellular. By upgrading to digital, however, cellular telephone carriers hope to match or beat upstart PCS operators in both price and performance.

The rush to develop a North American digital cellular standard has created a couple of unexpected problems. While the average consumer associates digital audio with noise-free, high-fidelity compact disks, the CTIA Interim Standard 54 (IS-54), also known as digital AMPS (D-AMPS), actually degrades voice quality. After all, the overriding concern was to increase capacity. In D-AMPS, voice is digitized at 8,000 bps (compared to 64,000 bps, the landline standard for digital voice). And consumers of the first D-AMPS services have complained of poor audio quality—particularly "warbling" tones, persistent background noise (which some liken to the sound of running water), and the fact that it is difficult to recognize familiar voices.

The first version of D-AMPS is also peculiarly hostile to mobile data. No provision is made for carrying data over the digital traffic channel. More amazingly, the traffic channel is incompatible with most modems. Early D-AMPS users must purposely revert to analog channels in order to transmit data. A future revision of the standard will, however, define a powerful array of circuit- and packet-switched mobile data services.

Of course, there is also plenty of good news. Although they are more expensive, handheld digital cell phones consume less battery power than their analog counterparts. Digital cellular provides greater privacy, since signals are unintelligible to mail-order FM scanners, and the bit stream is readily encrypted. Digital transmission will also enable cellular operators to implement quick and reliable end-to-end services such as automatic roaming and subscriber authentication. Indeed, most of the advantages PCS operators hoped to achieve can be realized in cellular's existing spectrum by upgrading to digital.

4.3 TIME, FREQUENCY, SPACE, AND CODE

The classic problem for the radio industry is how to accommodate as many users as possible in what appears to be a finite radio spectrum. The classic solution is frequency division multiple access (FDMA), where the spectrum is divided into various services (with associated block allocations) and each block is subdivided into channels which may be assigned to specific users or groups of users (Figure 4.1). FDMA may be used in conjunction with either analog or digital transmission.

Trunking was the first major improvement. Introduced in the 1970s, trunking permits a group of users to share a smaller number of channels on a contention basis. Instead of dedicating each channel to a single user and allowing the channel to lay idle when that assigned user is inactive, trunked radio systems juggle users around to make more efficient use of each channel. Trunking introduced the concept of time-sharing radio channels.

The next significant advance was cellular radio, which permitted the reuse of frequencies throughout a city. This is accomplished by dividing the service area into geographical cells and employing low-power transmitters. A more technical-sounding name for cellular radio is space-division multiple access (SDMA). (Figure 4.2).

These three techniques—FDMA, trunking, and cellular radio—may be implemented over analog radio systems. Time sharing, however, may be greatly enhanced by using it in conjunction with digital transmission. While trunking assigns a channel to a specific user for the duration of a communication session—often measured in minutes—it is possible for multiple users to simultaneously share a single channel if it continuously and rapidly switches from one to the next. In other words, the channel is divided into repeating frames consisting of a fixed number of time slots;

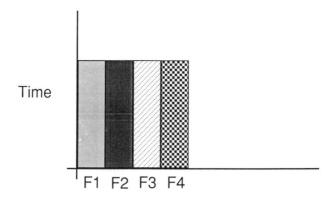

Figure 4.1 A frequency band divided into four channels allows four users to communicate at the same time (FDMA).

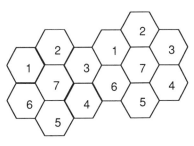

Figure 4.2 A geographical area may be divided into cells, among which seven groups of frequencies may be reused.

each user is assigned one or more time slots per frame. If the frames are repeated at a sufficiently rapid speed, each user experiences the illusion of operating on his or her own exclusive channel. This is called time division multiple access (TDMA) (Figure 4.3).

The last multiple-access technique we will consider here—CDMA—was introduced in the previous chapter. While each FDMA user is assigned a slice of frequency, and each TDMA user is assigned a slice of time, each CDMA user is assigned all of the available frequency and time. CDMA users are differentiated by their unique codes (Figure 4.4).

(The concept of code division may not be immediately obvious to our senses, but it is nonetheless valid. Just as a TDMA receiver can extract the desired user's transmit signal by listening during specific time slots and ignoring all others, a CDMA receiver can extract a desired user's transmit signal by listening for the user's specific code.)

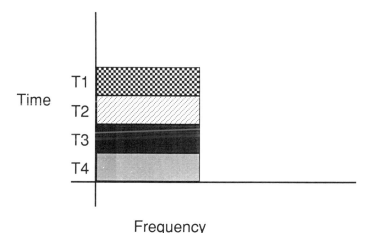

Figure 4.3 A single channel divided into four time slots allows four users to communicate at the same time (TDMA).

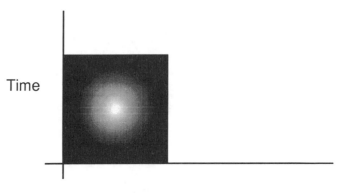

Frequency

Figure 4.4 In a CDMA system, users share all of the available frequency and time.

4.4 WHICH IS BEST: FDMA, TDMA, OR CDMA?

We have defined five different techniques for sharing a radio band, three of which have emerged as the leading digital radio contenders—FDMA, TDMA, and CDMA. Let's take a closer look at each.

Contrary to what you might expect, digital FDMA must consume more bandwidth than analog FDMA in order to convey the same audio quality. So why consider it? First, while analog FDMA requires two channels for full-duplex (simultaneous talk and listen) operation, digital FDMA can be combined with time-division duplexing (TDD) to support full-duplex operation over a single channel. Similar to TDMA, TDD divides the FDMA channel into alternating talk and listen time slots. Second, digital FDMA is more resistant to cochannel interference and should therefore enable frequencies to be reused more aggressively in cellular networks.

Now let us look at TDMA. A good example is D-AMPS, which supports three users per 30-kHz channel. The channel is divided into 25 frames per second, and each frame is divided into six equal time slots. Each user is assigned two of the six slots per frame—slot number 1 and 4, 2 and 5, or 3 and 6. Just as FDMA is often used in conjunction with TDD, D-AMPS uses TDMA in combination with frequency-division duplexing (FDD). In other words, separate frequencies are used for transmit and receive. For example, the user assigned to the first pair of slots will pick up bits in slots 1 and 4 on the receive channel, and drop off bits in slots 1 and 4 on the transmit channel. The transmit and receive frames are offset in time so that a user's transmit slots do not overlap his or her receive slots (Figure 4.5).

Three things are worth noting here. First, D-AMPS uses the same 30-kHz channels originally allocated for analog cellular. Second, D-AMPS yields an instant threefold capacity increase over AMPS. Third, D-AMPS is actually a TDMA/FDMA/ FDD system. (When it comes to spawning acronyms, the wireless industry is second to none!)

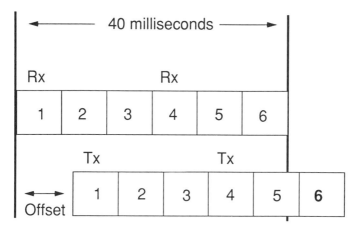

Figure 4.5 Transmit and receive frames are staggered in a full-duplex TDMA system.

Perhaps the most important advantage of TDMA over FDMA is that it requires fewer base station transceivers. For example, D-AMPS multiplexes the bit streams from three end users into a single base station transceiver for transmission, and demultiplexes the bit streams of three end users from a single base station receiver for reception.

But we cannot simply divide each channel into as many time slots as we please. For example, let us assume 8,000 bps is the minimum acceptable data rate for digitized voice. If the TDMA channel supports three users, then the aggregate channel rate must be at least 3 × 8,000 bps, or 24,000 bps. (For the sake of simplicity, we will ignore the additional throughput that must be reserved for error coding and control signaling in a real system.) The duration of each bit, therefore, is 1/24,000th of a second. Now let us increase the number of users to six. Each user still requires 8,000 bps, so the aggregate channel rate must now be 48,000 bps. Meanwhile, the duration of each bit has been cut in half: to 1/48,000th of a second. The shorter the duration of each bit, the greater likelihood it will be obliterated by an adjacent bit from a slightly delayed multipath signal. This is known as intersymbol interference (ISI). Because TDMA uses higher aggregate channel rates than FDMA, TDMA is more vulnerable to ISI.

Also note that on a TDMA channel serving three users, each user is idle for two of every three time slots. Of course, people do not speak in time slots, and each user's speech bits must be buffered and transmitted in bursts. These bursts are converted back into normally paced speech at the receiver. This buffer and burst process adds small end-to-end delays on top of the delays already incurred when converting (analog) speech into digital data, and digital data back into analog speech. Most two-way communications circuits are plagued by echoes—the speaker's voice is reflected back from the end of the link. If the echo returns in under 45 ms, it is not disturbing—and may actually serve as a pleasant confirmation that the communications link is functioning. But if the round-trip delay exceeds 45 ms—a condition more likely to occur with digital techniques such as TDMA—the echoes tend to become extremely annoying and must be suppressed.

Another difference between FDMA and TDMA is the way cell handoffs are handled. During a call, a (pure) FDMA handset is sending and receiving all of the time. To perform a handoff, the voice channel must be interrupted. A TDMA handset, on the other hand, is only sending and receiving a fraction of the time. It is possible to perform a handoff during a TDMA call without interrupting the voice channel, assuming the frames also carry signaling data.

(Although D-AMPS is a TDMA system, the voice channel must be interrupted during a handoff. The signaling bits in each frame—the slow associated control channel (SACCH)—lack the capacity to complete handoffs within the required time. Instead, handoffs are handled by the fast associated control channel (FACCH), which is created by momentarily seizing the voice channel. The FACCH, however, can complete a handoff in just 20 ms).

Now let's examine CDMA. In a fully loaded FDMA or TDMA system, users within a cell do not interfere with each other because they are segregated by frequency and (in the case of TDMA) by time. In a CDMA system, however, all of the users interfere with each other all of the time. Each additional user in a CDMA cell causes an incremental boost in the bit error rate (BER) for all users within that cell. In a single cell, therefore, CDMA is less efficient than either FDMA or TDMA.

The situation is very different, however, when one looks at the cellular network. Because FDMA and TDMA cannot tolerate cochannel interference, frequencies must not be reused in adjacent cells. Each cell, therefore, only uses a portion of the available bandwidth. In contrast, CDMA can tolerate cochannel interference, so adjacent cells may reuse all of the available bandwidth. In a cellular network, CDMA is more efficient than FDMA or TDMA.

While Table 4.1 illustrates some of the key differences between FDMA, digital FDMA, TDMA, and CDMA, it is not definitive. We are comparing apples to oranges to peaches to bananas. Vocoder rates, channel bandwidth, and sectors per cell are among the variables that define each implementation. The table, however, illustrates a couple of key points. First, digital FDMA can deliver twice the capacity of

Table 4.1
A Comparison of Analog FDMA, Digital FDMA, TDMA, and CDMA

	Analog FDMA	*FDMA*	*TDMA*	*CDMA*
Channels/MHz	30	30	5	1
Calls/MHz	15	30	20	12.5
Reuse pattern	7	7	3	1
Calls/MHz/cell	2	4	7	12.5
Capacity gain	1	2	3.5	6

analog FDMA simply by replacing a pair of frequencies (transmit and receive) with a single channel employing TDD. Second, while CDMA is less efficient than digital FDMA, TDMA, and even analog FDMA within an individual cell (calls/MHz), it is more efficient when deployed in a network (calls/MHz/cell).

CDMA compares favorably to FDMA and TDMA in other ways. Each CDMA base station requires just one transceiver. Because it uses a wideband signal, CDMA possesses inherent frequency diversity, and is more immune to multipath interference (see Chapter 1). (Remember, multipath interference is a frequency-specific effect, and CDMA signals contain redundant information spread over a range of frequencies.) Theoretically, FDMA and TDMA are simpler, but CDMA is more powerful.

4.5 DIGITAL VOICES

The human voice is naturally analog; that is, its sounds vary over a continuous range of amplitude (loudness) and frequency (pitch). Before we can transmit voice over digital radio, therefore, we must convert it into a stream of binary 1s and 0s. Digitized voice signals may then be compressed, multiplexed, or otherwise encoded for efficient communications—or encrypted for privacy.

In entertainment applications, voice is digitized to improve fidelity (by reducing noise) and ensure faithful reproduction. High fidelity requires high bandwidth for both recording and playback. For example, the new digital audio tape (DAT) standard runs at 768,000 bps. In digital radio, voice is digitized to increase capacity. This requires transmitting voice at the lowest possible bit rate—from 13,000 bps to as low as 2,000 bps.

There are essentially two ways to digitally encode voice signals. Waveform coding uses instantaneous measurements (or *samples*). The digitized signal may be transmitted to a receiver which reconstructs the original analog signal piece by piece. Parametric coding, on the other hand, matches sounds to stored speech models. Instructions are transmitted to the receiver, which assembles a signal that is an approximate reproduction of the original analog signal. While waveform coding is more accurate, parametric coding is more efficient (i.e., lower bit rates are possible using parametric coding).

The simplest form of waveform coding is pulse-coded modulation (PCM). PCM is widely used in landline telephony to digitize voice channels so that they may be combined (multiplexed) over high-speed digital circuits. For example, on a DS-0 channel (the U.S. standard), voice is sampled 8,000 times per second. Each sample is encoded as an 8-bit string (2^8 possible states). Therefore, a DS-0 channel runs 8,000 samples per second × 8 bits per sample, or 64,000 bits per second.

Parametric coding, on the other hand, looks at each voice waveform and encodes it as a set of values representing amplitude, pitch, and sound type. Sound types include *voiced* sounds (those that use the vocal chords), such as vowels, and unvoiced sounds (those that do not use the vocal chords), like the consonant *T*. Because parametric coding can achieve low bit rates without sacrificing intelligibility, it is a popular digital cellular solution. For example, full-rate D-AMPS channels use 8,000 bps parametric voice encoders/decoders (vocoders). It is expected that half-rate (4,000 bps) vocoders will be practical in the future.

4.6 THE GREAT MIGRATION

What was supposed to be a standard has turned into a free-for-all. The first North American digital cellular standard was TDMA-based IS-54. Qualcomm, however, convinced several carriers—AirTouch Cellular, US West New Vector Group, Alltel, and Sprint Cellular—that their CDMA-based technique is superior. Hughes Network Systems claims its extended TDMA (E-TDMA) matches the capacity improvement of CDMA, while offering backwards compatibility with the TDMA systems already being deployed. And proponents of a broadband CDMA scheme say their technology will allow carriers to overlay digital on their existing networks without (at least initially) sacrificing analog capacity.

But identifying a true standard is not the only issue clouding digital cellular deployment in the United States. The FCC has ruled that any upgrade to digital technology must not suddenly disenfranchise the millions of analog users. This means there will have to be transition a period, measured in years, during which there will be dual-mode networks and dual-mode phones. Of course, with competing digital techniques and uneven deployment, analog will remain the one true nationwide standard.

Perhaps the most important question is who is going to pay for the migration? If new subscribers are predominately low-volume users, they will be least inclined to foot the bill. Cellular carriers must provide incentives for high-volume business users to become the first digital converts. This means lower rates during peak rate hours.

Digital mobile (vehicular) phones were expected to start around $600 each, digital portables (handheld) at about $1,000. These products, however, are almost always bundled with a cellular service contract. In Chicago, for example, the Nokia 2120 D-AMPS/AMPS handheld (see Figure 4.6) was introduced at $200 along with a service contract. (Similarly, an analog phone can be bought for under $100 with a service contract.) Ultimately, digital-only phones will be cheaper than analog phones as further integration drives down prices. Cellular carriers are clearly willing to subsidize cell phones in hopes of winning long-term customers.

But network operators are also confronted with a disincentive: dual-mode networks are less efficient than single-mode networks. With the possible exception of broadband CDMA, digital channels can only be deployed by knocking out analog channels. Configuring just the right mix of analog and digital capacity is going to prove tricky, to say the least. If there are not enough digital channels, digital calls will be blocked and dropped at unacceptably high rates. If too many channels are

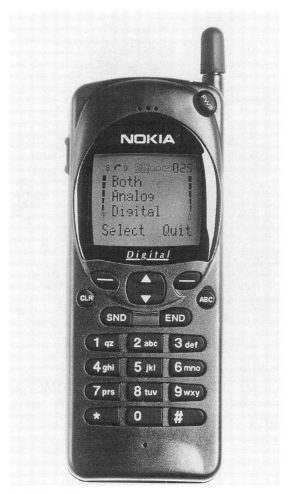

Figure 4.6 The Nokia 2120 TDMA phone.

allocated to digital, analog users will experience a precipitous decline in capacity. And most rural networks will probably delay upgrading to digital until urban networks start to phase out their last few analog channels.

4.7 NEXT-GENERATION STANDARDS

In Europe and Japan, government plays the lead role in setting telecommunication standards, and compliance is usually mandatory. In the United States, industry plays the lead role in setting standards, and compliance is generally voluntary. If the industry is unable to settle on a single standard, customers end up voting with their wallets. Let's take a closer look at the six next-generation cellular technologies

Table 4.2
A Comparison of N-AMPS, D-AMPS, E-TDMA, GSM, N-CDMA, and B-CDMA

	AMPS	*N-AMPS*	*D-AMPS*	*GSM*	*N-CDMA*	*B-CDMA*
Type	Conventional FM	Narrowband FM	TDMA/ FDMA	TDMA/ FDMA	Channelized CDMA	Broadband CDMA
Features	Voice	Voice + digital messaging service	Voice + CS, PS data	Voice + CS, PS data	Voice + CS, PS data	Voice + CS, PS data
Strengths	Widely deployed	3 × capacity of AMPS; paves way to digital	Fewer base station transceivers	Pan European network	Higher capacity than D-AMPS	Overlays existing AMPS users
Weaknesses	Insufficient capacity	Delays inevitable migration to digital	Must compete with CDMA	Requires migration to new band	Requires complex power control	Unproven technology
Status	Fully implemented	Implemented in spot markets	Implemented in spot markets	Widely deployed in Europe	To be deployed in 1995	Field demonstrated

CS = circuit-switched; PS = packet-switched.

vying to become de facto U.S. standards: narrowband AMPS (N-AMPS), D-AMPS, E-TDMA, narrowband CDMA (N-CDMA), broadband CDMA (B-CDMA), and the European export, GSM (see Table 4.2).

4.7.1 N-AMPS

If capacity is the main issue, it is not clear that digital is the best near-term solution. Motorola has developed a narrowband version of AMPS. Some cellular carriers (e.g., US West New Vector Group) see N-AMPS as a low-cost, low-risk means of relieving capacity bottlenecks and freeing up part of their spectrum for gradual conversion to digital.

By dividing each 30-kHz channel into three 10-kHz channels, N-AMPS provides a 3:1 capacity gain. The TIA has standardized N-AMPS in IS-88, IS-89, and IS-90. With its channels tightly packed together, N-AMPS requires better receiver selectivity (the ability to reject interference from nearby channels), which drives hardware costs up an estimated 15%. Like competing digital solutions, N-AMPS requires both new infrastructure and subscriber units; many carriers feel it makes more sense to go directly to digital. Carriers who have elected N-AMPS, however, are only converting a fraction of their channels.

Dubbed "advanced analog technology," N-AMPS is the first cellular service in the United States to offer digital messaging service (DMS), which supports end user text messages. Initial DMS capabilities are limited to one-way (network-to-subscriber) voice mail notification, callback number delivery, and transmission of "canned" messages. Ultimately, DMS will support two-way messaging (e.g., acknowledgment paging). DMS was derived from GSM's short-message service (SMS); comparable services are being defined for D-AMPS, CDMA, and even AMPS.

4.7.2 D-AMPS

TDMA also offers three times the capacity of analog, and will ultimately yield six times as much capacity through the use of half-rate (4,000 bps) vocoders. Digital also promises enhanced authentication, privacy, longer talk time for battery-powered portables, more consistent voice quality, and fewer dropped calls. With the rapid ascent of CDMA, however, it is by no means clear that D-AMPS (IS 54) will successfully establish itself as the North American digital cellular standard.

IS-54 was developed by the TIA TR45.3 committee. McCaw Cellular, Southwestern Bell, Bell Atlantic Mobile Systems, and BellSouth have committed to its deployment. While McCaw promised to introduce digital service in all of its markets by the end of 1993, the firm sees capacity as an urgent issue in only the largest markets. McCaw has also entered contracts to purchase digital phones from Ericsson-GE Mobile Communications, Hughes Network Systems, and Motorola.

TDMA enables one-channel-at-a-time conversion to digital. The first-generation dual-mode system uses the existing control channels; that is, previously reserved bits are used to request and assign digital or analog voice channels. (Later revisions will include digital-only control channels.) Whenever a mobile originates a call or responds to a page notifying it of an incoming call, it indicates its preference (analog or TDMA) over the reverse (mobile-to-base) control channel (RCC). The cellular network then assigns either an analog or TDMA channel over the forward control channel (FCC).

IS-54 uses what is known as vector-sum excited linear predictive coding (VSELP) to digitize speech at 7,950 bps. This technique distorts musical tones. While merely annoying when listening to music on hold, it is problematic for dual-tone multifrequency (DTMF) response systems like voice mail, fax request services, and remote control answering machines (i.e., systems that respond to familiar telephone keypad tones). The solution will be to install DTMF tone generators at the MTSO; whenever a subscriber depresses a keypad button, a special command will be sent to the MTSO to generate the appropriate tone.

More important, vocoder distortion will obstruct the use of modems over digital voice channels. Since the first version of D-AMPS being implemented (IS-54 Rev. B) does not define data services (i.e., standardized ways to circumvent the vocoder), an ironic situation has arisen: digital cellular is more hostile to end-user data than analog cellular. The initial D-AMPS data solution is that the mobile unit must request an analog channel for use with a modem. Two steps forward, one step backward!

4.7.3 E-TDMA

The biggest challenge facing D-AMPS is how to fend off the higher capacity promises of CDMA. Hughes Network Systems (Germantown, Maryland) has developed E-TDMA, which serves on average six voice conversations for every 30-kHz channel—effectively doubling the capacity of D-AMPS.

The secret to E-TDMA's capacity is digital speech interpolation (DSI), a system that takes advantage of the silent intervals that comprise a large portion of the time during ordinary conversations. For example, ten voice conversations may simultaneously share eight channels if bandwidth is allocated only to those users actually speaking at each instant. This requires a technique known as voice activity detection (VAD), in which speech spurts are detected and assigned a frequency/time slot with imperceptible delay (within a few milliseconds).

There are additional benefits to E-TDMA. The mobile phone only transmits when the user is speaking, conserving battery power. E-TDMA also has a "soft" capacity limit: the maximum number of simultaneous users is not fixed. However, as the number of users increases, so does the risk of losing occasional voice spurts. E-TDMA is backwards-compatible and can coexist with D-AMPS. Hughes is working to make E-TDMA an official standard and is also promoting the technology for wireless local loop applications in areas like the former Soviet Union and Eastern Europe, where the telecommunications infrastructure is badly outdated. The Hughes GMH 2000 digital cellular system can be software-configured for AMPS, TDMA, or E-TDMA (and also cellular digital packet data; see Chapter 5).

4.7.4 Global System for Mobile Communication

Conceived in the early 1980s, GSM originally stood for *Groupe Special Mobile*. Its goal is to provide a pan-European cellular telephone system offering cross-border compatibility, greater economy of scale for manufacturers, and ultimately competition for local telephone services. As of August 1994, 32 operators in 22 countries (several outside of Europe) are deploying GSM networks. GSM uses the same frequencies (890- to 915-MHz mobile transmit; 935- to 960-MHz base station transmit) in 12 member states of the Commission of the European Communities (CEC).

GSM is a committee-defined solution described in over 15,000 pages of documentation. Not surprisingly, GSM has done its share in expanding the list of wireless acronyms (Table 4.3).

GSM promises three times the capacity of present analog systems, plus a variety of new services such as international roaming, compatibility with landline ISDN service, international ISDN phone numbers, and enhanced authentication.

Table 4.3
Selected GSM Acronyms

Base station subsystem	BSS
Mobile switching center	MSC
Home location register	HLR
Visiting location register	VLR
Service control points	SCP
Equipment identity register	EIR
Authentication center	AC
Mobile application part	MAP
Mobile station ISDN	MSISDN
Mobile station roaming number	MSRN
Traffic channel	TCH
Slow associated control channel	SACCH
Fast associated control channel	FASSCH
Broadcast control channel	BCCH
Synchronization channel	SCH
Frequency control channel	FCCH

Voice, fax, data, and paging services are supported. Like D-AMPS, GSM is a TDMA/FDMA/FDD system. The two 25-MHz-wide GSM bands are composed of 125 channels, each of which are 200 kHz wide. GSM was designed to support up to five carriers, allocating 10 MHz to each (3 to 6 channels per cell per carrier).

Each channel is divided into eight time slots (one per user); transmit and receive frames are staggered by three time slots. User devices transmit at 16 kbps, of which approximately 13 kbps is dedicated to voice and 3 kbps to channel signaling. Transmission occurs in bursts lasting 0.546 µs—slightly shorter than the 0.577-µs time slots in order to accommodate delay dispersion and timing errors.

While CDMA has been harshly criticized for its near-far problem (see Chapter 3), it turns out that TDMA has its own near-far snag. In D-AMPS, the duration of each transmit symbol is 41 µs—the time it takes a signal moving at the speed of light to traverse approximately 7.6 miles. This is not a serious problem in D-AMPS, because the typical cell radius is much smaller than 7.6 miles. But in GSM, transmit symbols have a duration of just 3.6 µs—the time it takes for a radio signal to travel about two-thirds of a mile (approximately 1.1 km). With GSM, there is a significant chance that signals from mobiles near the edge of a cell will spill over into the time slots of mobiles close to the cell site. In fact, GSM was designed to work with cells ranging from 200m to 35 km in radius (the latter demanded by Scandinavian countries).

Just as dynamic power control is required to defeat CDMA's near-far problem, *dynamic time alignment control* is required to overcome GSM's near-far problem. This is accomplished by requiring the mobile station to echo back synchronization bits transmitted by the base station along with its mobile identification number. The base station continually monitors the propagation delay for each active mobile; mobile transmitters are brought into alignment by assigning appropriate time delays to each.

Of course, the problem is more complicated. What happens when a mobile is powered up and makes its first transmission? Or what if a mobile has just been handed off from one cell to another? The mobile must make at least one transmission for the base station to make its first measurement. To prevent the first transmission from trampling that of another mobile, a special short burst (carrying no speech bits) is sent. And in networks with particularly large cells, such as in rural Scandinavia, mobile transmissions are assigned to every other time slot only—eliminating interference caused by time-slot spill over, but also reducing system capacity by one-half.

GSM has another serious technical problem: GSM handsets cause interference to nearby electronic devices, most notably hearing aids and heart pacemakers. This is not due to incompetent design or shoddy construction on the part of individual manufacturers, but is inherent in GSM's TDMA scheme. The GSM mobile device (unlike the base station) transmits only during assigned time slots; when the user speaks, the transmitter is rapidly switched off and on. This constant transmitter toggling generates low-frequency signals at 217 Hz, 434 Hz, and 651 Hz—the last two creating low-pitched buzzing sounds in nearby hearing aids. (In the United States, D-AMPS handsets generate signals at 100, 200, and 300 Hz, which may also cause some interference. In Australia, GSM handsets have produced interference in the adjacent analog cellular band.)

There is only one known cure for GSM's interference problems: reduce handset transmit power. This can be accomplished by constructing microcells in densely populated areas (something carriers were likely to do, anyway). Larger cells and higher power levels can be used in more sparsely populated rural areas.

(GSM proponents generally blame the problem on manufacturers of concerned devices, such as hearing aids. Although such devices were on the market first and were long deemed acceptable, we must remember that these manufacturers are up against a government-sanctioned GSM industry. Hearing aid manufacturers will probably be forced to redesign their products.)

GSM is a truly "personal" service in that it supports both network and terminal portability. Subscribers can use any device on any network by simply inserting their subscriber identity module (SIM), or smart card, in the device. Voice is digitized using RPE-LPC (regular pulse excited, long-term prediction) vocoders. In the future, VAD will be added to reduce portable power consumption, and half-rate vocoders will be employed to double capacity. A derivative of GSM known as DCS-1800 (Digital Cellular System for 1,800 MHz) has been defined for personal communications networks (PCN).

The rollout and acceptance of GSM has been somewhat less rapid than had been expected. The protectionist tendencies of some countries has been a problem. The GSM Memorandum of Understanding (MoU) was signed by 17 European countries and 8 countries outside Europe. The signatories, however, had to promise to deploy a minimum number of base stations—particularly at airports, along highways, and in city centers—to ensure international roaming. GSM handsets have been somewhat slow to come to market due to cumbersome type-approval procedures still in force in certain countries. And when Motorola entered the GSM market with a relatively low-priced handset, some accused the firm of "dumping." Nevertheless, GSM represents the first major (if somewhat flawed) deployment of digital cellular technology and is already demonstrating its superiority over analog cellular.

4.7.5 N-CDMA

Qualcomm has argued that D-AMPS is not a sufficiently powerful solution for digital cellular, and they have convinced a significant portion of the industry: the TIA TR45.5 committee approved CDMA as IS-95 in the summer of 1993. While D-AMPS promises a threefold capacity gain, N-CDMA claims it can deliver 10 to 15 times the capacity of AMPS. Through its advocacy of CDMA, Qualcomm has won adherents among network operators, equipment manufacturers, and Wall Street investment analysts. Motorola, the world's largest wireless communications equipment manufacturer, is building CDMA systems in the United States, the Philippines, and Hong Kong. Firms developing CDMA subscriber devices include OKI Telecomm, Motorola, Qualcomm (Figure 4.7), and Nokia.

Qualcomm's CDMA channels consume 1.8 MHz of bandwidth: a 1.25-MHz voice channel surrounded by two 275-kHz guard bands. Each CDMA channel can accommodate 25 to 40 active calls. Therefore, ten channels can accommodate 250 to 400 simultaneous calls per cell. (A comparable D-AMPS system would support about 175 calls per cell.)

Qualcomm claims it has developed the necessary dynamic power control technology. The three main factors that contribute to signal strength variations are distance, shadow fading (caused by obstacles), and Rayleigh fading (multipath interference). Brief fades can be dealt with through the use of coding techniques such as forward error correction (FEC); that is, the receiver can reconstruct short strings of bits lost to Rayleigh fading. While coding could also be used to combat longer fades—bits can be interleaved to enhance the effectiveness of FEC—this would introduce unacceptable delays for voice applications. Power control, therefore, must be employed to deal with distance-dependent and shadow fading.

Signals received at the cell site must be adjusted to within 1 dB of each other in strength. (Decibels are used to measure the power ratio between different signals.) Although difficult to implement, power control makes a virtue out of necessity by ensuring that the mobile device always transmits with the minimum power. Although other cellular systems also use power control, its action is (necessarily) much more rapid in a CDMA network (Figure 4.8).

N-CDMA offers other advantages including "soft" cell handoffs, graceful ca-

Figure 4.7 A Qualcomm CDMA phone.

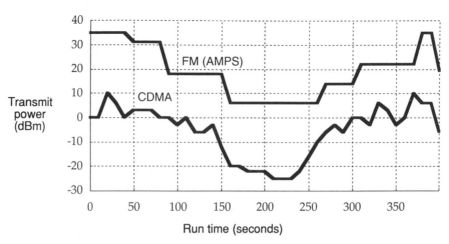

Figure 4.8 Qualcomm's dynamic power control is complex, but allows the cell phone to operate at significantly lower transmit power levels.

pacity degradation, the use of *Rake* receivers to combat multipath propagation, and an inherent form of DSI. Soft handoffs occur when a call temporarily exists on two adjacent base stations. This is possible because both base stations operate on the same frequencies. Soft handoffs minimize the risk of dropping calls due to premature or late handoff. When a mobile user is at a cell boundary, the best signal at each instant dominates. Qualcomm also claims to have developed softer handoffs in which a call is handed off between sectors within the same cell.

Like E-TDMA, there is no hard limit to CDMA's capacity. Every time an additional user appears on the CDMA channel, all of the users experience an incremental rise in background noise. As signal quality gradually degrades, some users will decide to hang up and try again later. In this way, Qualcomm believes that commercial CDMA networks will be self-regulating.

Rake receivers are used to combat multipath propagation. The Rake receiver locks on to a number of multipath signals, adds appropriate delays to the early arriving signals so that all are in phase, and then recombines them. In other words, by manipulating multipath signals, it is possible to ensure that they reinforce rather than interfere with each other. Some industry experts, however, question whether Rake receivers are practical (i.e., can be achieved at an acceptable price/performance ratio). The Rake receiver is implemented on a single chip in the mobile unit and on multiple chips in the base station. According to Qualcomm's Andrew Viterbi, once a working Rake receiver is embedded in silicon it will become a standard building block and will come down in price regardless of its complexity.

While E-TDMA uses DSI to increase capacity, it does so at the risk of losing parts of voice transmissions, since occasionally there may be no vacant slots. Qualcomm, in contrast, employs VAD to control transmission bit rate. An N-CDMA voice path may operate at 8, 4, 2, or 1 kbps. (It never falls below 1 kbps, however, as this is the minimum bit rate required to keep dynamic power control functioning.) Leveraging VAD, N-CDMA voice paths operate at an average rate of 4 kbps.

One strategy for implementing CDMA is to make it the final stage in an upgrade process that begins with N-AMPS. For example, US West New Vector Group intends to divide its cellular telephone spectrum into 12.5 MHz for AMPS, 10 MHz for CDMA, and 2.5 MHz for N-AMPS, with CDMA in the middle. A number of carriers have field-tested CDMA and plan to offer commercial service by the end of 1995. AirTouch Cellular says it has verified Qualcomm's claim that CDMA delivers 12 × 15 times the capacity of analog AMPS. Bell Atlantic Mobile Systems (BAMS) found that CDMA calls could, on average, be maintained at signal levels 5 dB below those of analog FM. In Ameritech's field trial of D-AMPS and CDMA, most users preferred the audio quality of CDMA. However, it is important to note that BAMS has chosen D-AMPS for certain markets, while Ameritech decided to defer its decision to a later date. It is a testimony, however, to the compelling nature of CDMA (and no doubt Qualcomm's persuasiveness) that progeny of the RBOCs have seriously considered— let alone selected—CDMA as their digital cellular solution.

Qualcomm is aggressively promoting CDMA for the new PCS in the United

States. Spectrum will not come cheaply to PCS operators, who will have to pay for both their licenses and the relocation of incumbent microwave users, a fact that Qualcomm hopes will propel operators to select the highest capacity technology. The firm is also working on the development of SMS and mobile data standards for CDMA and on promoting the development of compatible wireless PBX and residential cordless telephone base stations.

4.7.6 B-CDMA

InterDigital Communications has proposed a broadband CDMA (also known as wideband CDMA) solution for digital cellular. Because it is designed to coexist with narrowband users, B-CDMA would enable cellular telephone carriers to upgrade to digital without having to immediately sacrifice analog capacity. Given resistance to the two-way battle already raging (D-AMPS versus N-CDMA) and InterDigital's lack of third-party support, it seems unlikely that B-CDMA will emerge as a serious contender in the cellular telephone industry. But B-CDMA is a higher-performance version of Qualcomm's technology and could certainly become a second-generation PCS technology—perhaps for deployment early in the 21st century.

InterDigital became a B-CDMA proponent when it acquired SCS Mobilcomm, a firm headed by spread spectrum expert Donald Schilling. Schilling, along with Shelby Bryan, helped launch PCS in the United States when they applied for and received an experimental license to conduct tests in the 2-GHz band. Schilling believes PCS will succeed because it will enable one phone to operate over both private (mainly indoor) and public (mainly outdoor) systems.

Schilling points out that broadband is more robust than narrowband spreading. B-CDMA possesses greater frequency diversity than N-CDMA, and is therefore more immune to multipath interference (and does not require Rake receivers). This is particularly important for indoor applications, where path length differences are generally small. An N-CDMA wireless PBX, for example, must deliberately create multipath signals for the Rake receiver. This is done by using a distributed antenna system (i.e., antennas located at intervals along a cable) and by injecting artificial delays. Like N-CDMA, however, B-CDMA must also employ dynamic power control. (Schilling's B-CDMA makes adjustments every 250 μs in steps as small as 0.25 dB.)

How does B-CDMA coexist with analog cellular? Schilling illustrates the concept this way: Imagine spreading a 1W signal over 100 MHz. To a user listening on a 1-MHz-wide channel, the signal looks like a 10-mW interferer. The main problem is not interference caused by B-CDMA transmitters to narrowband analog receivers, but the interference caused by multiple (and relatively powerful) analog transmitters to B-CDMA receivers. The solution devised by Schilling is a series of adaptively adjustable notch filters (devices that attenuate signals over a narrow frequency range). The B-CDMA system monitors analog channel activity (easily done, since their base stations are collocated at the cell sites) and adjusts the mobile receivers' notch filters accordingly.

4.8 DIGITAL SMR NETWORKS

SMR operators—the providers of "blue collar" two-way radio services—are also journeying to the brave new world of digital cellular. But in adopting a digital cellular architecture, they have stirred up quite a controversy. By dramatically increasing their capacity, SMR operators could beat PCS licensees to the punch in becoming the U.S.'s third cellular player.

When Rutherford, New Jersey–based Nextel (previously known as Fleet Call) obtained FCC permission to build a digital cellular network, the decision was met with a hailstorm of criticism. Most vocal were the cellular telephone subsidiaries of the RBOCs. The RBOCs' main complaint, however, was not that the petition should have been denied, but that restrictions on cellular carriers should also be lifted. (At the time, SMRs were classified as unregulated *shared private* carriers, while cellular carriers were treated as regulated common carriers. Ironically, cellular carriers were hoping to be treated more like SMRs; instead, the FCC is treating SMRs more like cellular carriers. See Chapter 2.)

There are actually two competing digital SMR architectures: digital SMR (DSMR) and enhanced SMR (ESMR). DSMR simply divides each 25-kHz channel into three (or six) voice channels. Actual capacity increase would be somewhat greater due to improved trunking efficiency. (With proportionately fewer channels and users, channels tend to remain idle for longer periods, resulting in lower efficiency.)

Prior to 1990, SMRs were not permitted to construct their networks in cellular fashion. Nextel's petition asked the FCC to allow it to construct ESMR networks based on a cellular TDMA technology developed by Motorola called MIRS (Motorola Integrated Radio System). Each 25-kHz SMR channel is divided into six time slots. With a sixfold capacity increase due to TDMA, and a fivefold increase due to cellularization, an overall thirtyfold increase in system capacity can be realized.

Both DSMR and ESMR may find a place in the SMR business. ESMR will probably be used in large cities, and DSMR (noncellular) in small- to medium-sized cities. Motorola points out that it will be less expensive and easier to migrate to DSMR. ESMR will also support roaming (not supported by most of today's SMR services). Other features offered by MIRS include automatic forwarding of mobile ID to dispatcher when the user presses the "push to talk" (PTT) button; enhanced PSTN interconnect capabilities like call waiting, three party calling, call forwarding, and outbound dialing restrictions; SMS; and both circuit-switched and packet-switched mobile data.

Nextel's Chairman Morgan O'Brien is legendary for his deal-making abilities.

The firm has acquired hundreds of SMR licenses from Motorola, merged with other SMR operators, and raised over $1 billion for the construction of its network. At the same time, Motorola—once the largest SMR operator in the United States—has managed to exit the service business in order to focus on manufacturing and selling ESMR infrastructure and mobile equipment.

A wild card in the digital SMR push is New Jersey–based Geotek. The firm has developed an alternative digital SMR based on spread spectrum technology developed in Israel by Rafael, the Israeli Ministry of Defense's R&D agency. Its FHMA approach provides voice and data services via 40 narrowband (12.5-kHz) channels. Geotek claims its technique will deliver 30 times the capacity of analog SMR, while costing about one- eighth of a comparable TDMA network (covering the area within a 35-mile radius). The key to FHMA's low cost and high capacity is the fact that it uses one large sectorized cell instead of many smaller cells. Hopping patterns are coordinated to prevent interference between users.

Because it relies on large cells, Geotek's system is only suitable for vehicular users. Given the fact that Nextel appears to be focused on competing with cellular telephone, and its Motorola MIRS technology is relatively expensive, there may be an opportunity for Geotek to continue to expand blue collar markets in which dispatching, rather than mobile telephony, is the major application.

4.9 "INSURMOUNTABLE OPPORTUNITY"

Digital cellular technology will make mobile communication services accessible to the majority of people in the developed countries. It will do this not by being miserly with bandwidth, but by taming the interference beast. The wavelengths that course through copper, coaxial, and fiber-optic cables will also conquer free space—leveraging microprocessors, spreading codes, and microcells along the way.

Many observers believe Europe is ahead of the United States in the development of mobile communication technologies, primarily thanks to the establishment of the GSM digital cellular standard and progress towards its implementation. Nothing could be further from the truth. While GSM may solve some of the problems that have plagued Europe's cellular industry, it does not represent a major technological advance. Developed by committee, wrought through compromise between corporate and government leviathans, and implemented by decree, GSM serves entrenched interests. Those who counsel the United States to embrace European- style industrial policy forget that pivotal computer and communications innovations—from the microprocessor to cellular telephone—originated under the regime they would have us abandon.

In the immortal words of Pogo the cartoon character, "We are surrounded by insurmountable opportunity." In the United States, there are many competing digital cellular technologies and there are many competing business plans. It is confusing and messy. But this is the definition of opportunity—and it is only made insurmountable when we begin to limit our choices.

SELECT BIBLIOGRAPHY

[1] Arnbak, J. C., "The European (R)Evolution of Wireless Digital Networks," *IEEE Communications*, April 1991.

[2] Calhoun, G., *Digital Cellular Radio*, Norwood, MA: Artech House, 1988.

[3] Fist, S., "GSM—Better Late Than Never?" *Australian Communications*, June 1993.

[4] Fist, S., "Grand Scale Mistake?" *Australian Communications*, July 1993.

[5] Fist, S., "Will GSM and D-AMPS Give Way to the CDMA Push?" *Australian Communications*, July 1993.

[6] Holtzman, J., Rutgers University WINLAB, interview on August 10, 1993.

[7] Malgieri, S., "Engineering Committee OKs CDMA as Standard," *Radio Communications Report*, Denver, CO, August 9, 1993.

[8] "McCaw Deal Positions AT&T to Offer PCS," *MicroCell News*, Probe Research, Inc., Cedar Knolls, NJ, August 25, 1993.

[9] Presentations from CDMA Digital Cellular Technology Forum, San Diego, California, February 23–24, 1993.

[10] Rahnema, M., "Overview of the GSM System and Protocol Architecture," *IEEE Communications*, April 1993.

[11] Reijonen, P., "GSM Base Station Development," *Telecommunications*, Sept. 1990.

[12] Schilling, D., InterDigital Communications Corporation, interview on August 11, 1993.

[13] Smolik, K., "U.S. Standards: Digital Cellular," AT&T Bell Laboratories, presentation at Chicago IEEE Wireless Communications Symposium, October 29, 1992.

[14] Tuttlebee, W. H.W., ed., *Cordless Telecommunications in Europe*, London: Springer-Verlag, 1990.

[15] Varrall, G., and R. Belcher, *Data Over Radio*, Mendocino, CA: Quantum Publishing, 1992.

[16] Viterbi, A., Qualcomm, Inc., interview on August 13, 1993.

CHAPTER 5
▼▼▼

MOBILE DATA

5.1 WHY MOBILE DATA WILL SUCCEED

Although still nascent, mobile data has begun to demonstrate compelling business applications. Mobile data is being used to help organizations manage their personnel and assets in the field, enhance the productivity of field workers, and—most important—provide faster customer response. It is the last of these applications that will most likely propel mobile data into the mainstream. As the economies of developed countries grow increasingly service-oriented, the ability to reach and serve customers becomes crucial as vendors battle for market share.

In business, mobile data promises to extend office automation into the field. For years, field workers have managed to avoid using the computers that have overrun our offices. This is not surprising, since the people who work in the field—truck drivers, field service technicians, police officers, sales representatives, and others—are not traditional knowledge workers. But while the "paperless office" never materialized—paper is the lifeblood of knowledge workers—there are convincing reasons to eliminate paper in the field.

Wireless tablet computers will replace the multipart forms that often fall victim to coffee spills and inclement weather. Using electronic forms with signature capture, sales orders will be instantly entered, verified, and relayed to factories. The additional step of entering handwritten data into a computer system—so prone to transcription errors—will be eliminated. Ultimately, computers will enable field workers to spend more time doing what they were hired to do—like selling, repairing, or delivering.

Mobile data will set new business standards for timely access to people and information. Managers, business planners, and account executives—all of whom are expected to spend more of their time in the field—will profit from remote access to enterprise networks. Workgroups will share appointment calendars, databases, and team-developed documents. Peer-to-peer messaging will be precisely that, once users are sprung loose from their desktop tethers.

Ultimately, mobile data will find use among consumers. The possibilities are endless. People have begun to use pagers and pay phones to coordinate family activities; two-way wireless messaging devices will grant them instant access and response. We may one day do away with road signs, replacing them with dashboard-mounted *virtual signs*. Tourists will use mobile data to access information about historic sites and nearby services and shops, and as a navigation aid.

But mobile data development is still very much in its infancy. Market researchers have predicted the number of mobile data users will grow to as many as 13 million by early next decade. By late 1994, however, there were only about 350,000 mobile data users in the United States. Nevertheless, the industry is growing at a healthy clip, and a dazzling array of new products and services on the way could create a huge surge in mobile data activity.

5.2 MOBILE DATA APPLICATIONS

There are many different types of mobile data requirements. They may be classified as to whether they are primarily urban or rural, in the vehicle or on foot, one-way or two-way, interactive or store-and-forward, data-only or voice plus data, wide area or metropolitan area, and whether they require radiolocation or not. But from the user's perspective, there are two major categories of applications: field automation and personal messaging.

The best known field automation application is computer-aided dispatching (CAD). Dispatching centers manage transportation, freight, and field service operations. The dispatcher tells the driver or technician where to go and when; the field worker reports status and requests further instructions.

Most dispatching is still accomplished the inefficient way—via voice communications. Time must be spent establishing contact, acknowledging instructions, and transcribing information. An amazing amount of time may be consumed just tracking down a telephone. And the field worker must remain within earshot of a vehicular voice radio in order to receive messages.

With CAD, all of the protocol functions are performed automatically. It is not necessary for the driver to be present to receive a message. The time required to exchange messages is reduced. Errors caused by relaying information verbally are eliminated. The number of dispatchers required is greatly reduced. And because the field worker has direct access to a computer, new services may be added by simply replacing or upgrading the software.

CAD is well established in the long-haul trucking industry. The ability to communicate while on the road eliminates the need to make "phone check" stops. More importantly, dispatchers can minimize "out-of-route" and "empty" miles by continuously tracking vehicles; when a new pickup order is received, a nearby or approaching truck can be instantly redirected. Real-time communications also has safety benefits: the status of trucks carrying hazardous materials may be monitored, and drivers can signal for help in the event of an emergency.

Celadon Trucking Services is equipping 1,000 trucks with mobile communications terminals for use over Qualcomm's OmniTRACS satellite-based data network. After only the first 170 units were installed, the firm found its telephone charges had dropped by $30,000 per month. According to Celadon, wide-area mobile data service saves time for drivers, enables dispatchers to be more proactive, and provides customers with a greater sense of security—the exact whereabouts of their shipments are known at all times.

Taxicab firms in major cities such as Seattle, Washington, and London, England, have embraced CAD. A dashboard-mounted terminal is used to browse and select open orders, record fares, or even download navigational information. Orders are listed by geographical sector, enabling drivers to quickly identify and secure nearby pickups. But this application also demonstrates that not all field workers welcome automation. A taxicab company in Texas abandoned its CAD project when drivers complained the computerized system was "too fair"; it eliminated the advantages senior cabbies had acquired over the years through personal contact with dispatchers!

Field service organizations are using mobile data for dispatching, flagging contract requirements, checking parts inventory, ordering parts, and even providing access to expert systems. The last item—remote access to expert systems—has tremendous promise for reducing operating costs, because it enables field service organizations to employ less highly skilled and therefore less highly paid field technicians.

Firms such as postage meter manufacturer Pitney Bowes have conducted time and motion studies revealing that field service technicians spend, on average, over an hour per day searching for and using telephones. (It is not that there are not enough telephones—it is finding ones that are available.) It is also difficult for a dispatch center to originate communications via telephone with a technician in the field.

(In the case of Pitney Bowes, they were interested in substituting wireless data access for dial-up data communications via acoustic couplers. They immediately recognized that in order to realize a cost savings, a wireless network had to provide 90% coverage. But they discovered that coverage varied depending on whether the technician was standing next to the customer's equipment, a window, or just outside the building. They also realized that with anything less than 100% coverage, it would be necessary to maintain dual connectivity, so that technicians could still communicate from locations not covered by the wireless network. Pitney Bowes eventually selected ARDIS, described below.)

Another compelling mobile data application—still largely untapped—is sales force automation. Wireless data networks will enable field sales professionals to download additional product information, check inventory, enter orders (with an "electronic carbon copy" whisked directly to the factory), and confirm delivery dates—all while sitting in front of the customer. Some organizations are evaluating mobile data as a means of streamlining sales forecasting and reporting functions. The goal is to keep field sales professionals out in the field—where customers are— instead of back at the office filling out reports. Along with field sales representatives, consultants and financial auditors are also being pressured to stay in the field, forcing members of these professions to establish *virtual offices* in their briefcases— often replete with battery-powered printers, fax machines, and wireless access to the enterprise network.

In a related application, ADP Automotive Claims Services is using wireless data communications to enable insurance claims adjusters to access comprehensive, frequently updated databases to assist them in preparing accurate repair estimates and speeding insurance claim processing. In this way, the claims adjuster can prepare an estimate directly from the auto repair shop's garage or parking lot.

Courier services such as Federal Express and United Parcel Service (UPS) in the United States and Parcelforce in the United Kingdom, are among the largest mobile data users. Federal Express's private nationwide wireless data network, built by Motorola, enables the firm to track packages from pickup to delivery using bar code identification at each point along the way. Federal Express's success forced UPS to deploy a wireless package tracking solution of its own, albeit nearly a decade later. UPS's implementation (like Parcelforce in the United Kingdom) uses modems over analog cellular; the firm offers major customers the option of directly accessing their computer network for immediate access to package status.

There are numerous applications for mobile data in public safety. Firefighters are using wireless data communications to retrieve information on burning or threatened commercial buildings, such as blueprints showing locations of ventilation ducts, sprinklers, and highly flammable materials. Hazardous material (HAZMAT) swat teams are accessing extensive chemical databases to determine how best to clean up highway spills or illegal dumps. Mobile data can also be used to dispatch fire trucks, provide navigation information, and record and time stamp-events for later review (e.g., how long did it take to arrive on site?).

Mobile data links may be used to upload digitized x-rays or electrocardio-grams from accident sites so that physicians waiting at nearby hospitals can give paramedics special instructions. It is now possible for police to download hotsheets containing photographs of wanted criminals. U.K.-based Origin has developed a system for compressing and transmitting still images over low-speed mobile data networks. The firm claims it can reduce a 300K image down to about 10K, making it possible to transmit a mug shot over RAM Mobile Data's Mobitex network for about $2.50.

Wireless electronic mail and two-way peer-to-peer messaging could become mobile data's "killer applications." These applications represent a convergence between the paging, portable computer, and electronic mail industries. The ability to both send and receive messages is an obvious enhancement over today's highly popular paging services. The ability to extend e-mail access beyond the desktop—to meeting rooms, the cafeteria, colleagues' offices, and even on the road—can only enhance the power of e-mail. According to a study by the Electronic Mail Association (EMA), there may be as many as 25 million e-mail users, transmitting over 15 billion messages annually, in the United States by 1995. There are over 7 million business travelers in the United States who could potentially benefit from wireless e-mail, transforming it into the medium of choice for coordinating appointment calendars, obtaining multiple approvals, and communicating with colleagues.

5.3 DATA OVER ANALOG CELLULAR

With its massive infrastructure, financial muscle, and existing subscriber base (19,000,000 users in the United States alone as of late 1994), many believe cellular telephone is the best platform for mobile data. And data over analog cellular is particularly attractive; enhanced versions of standard dial-up modems, along with familiar PC or Macintosh communications software, can be used to access the same services available to desktop users.

But data transmission over today's analog cellular networks is fraught with challenges. First, most cell phones do not include an interface for auxiliary equipment like modems or fax machines. Several firms manufacture adapters that insert "Y" cables between the handset and radio unit of two-piece vehicular and transportable phones, or plug directly into proprietary jacks on handheld phones, to provide RJ-11 connections for standard telephone equipment. One of the first such devices was the CelJack manufactured by Telular (Buffalo Grove, Illinois) (Figure 5.1).

There are both "intelligent" and "dumb" RJ-11 adapters. Intelligent adapters behave much like plain old telephone service (POTS) lines: they present dial tone, accept DTMF tone dialing, generate ring signals, and go on-hook and off-hook. Dumb RJ-11 adapters simply convert the handset's four-wire connection to a standard telephone line's two-wire connection; the user must manually originate and answer calls. The first RJ-11 adapters were bulky and designed to be powered by a vehicle's battery. Newer units are pocket-sized and operate off internal 9V batteries, opening the door to widespread use of data over cellular.

The analog cellular voice channel also throws up a number of obstacles to data communications. Because AMPS voice channels are slightly narrower than landline voice channels, modems designed to operate at 9,600 bps (full duplex) and above must often fall back to 4,800 bps. During cell handoffs, audio is interrupted for 0.2 to 1.2 sec, ordinarily causing dial-up modems to hang up. This can be prevented by resetting the modem's loss-of-carrier time-out to 2 sec or greater (using industry-

Figure 5.1 Telular's CelJack enables auto-dial modems to be interfaced with mobile and transportable cell phones.

standard AT commands). Although handoffs are a normal part of mobile operation, they may also occur with a stationary cell phone, because cellular networks sometimes redistribute traffic from heavily loaded cells to less heavily loaded adjacent cells. Finally, cellular voice channels are susceptible to all of the impairments endemic to mobile radio, such as cochannel interference and multipath fading, which may cause the loss or corruption of data.

Channel impairments may disturb data-over-cellular data communications in two ways: (1) they may disrupt the initial modem handshake causing the modems to either abort the call or establish a lower speed connection, and (2) they may cause bit errors during a data session. A number of special protocols have been developed to help deal with these problems. One drawback to such protocols, however, is that they usually require a compatible modem at the other end of the link.

(Some cellular carriers have installed modem pools in their MTSOs. The mobile subscriber dials a special number to access a cellular protocol modem; this modem is interfaced back to back with a Hayes®-compatible modem. Once a connection to the cellular modem is established, the user can dial out through the Hayes-compatible modem to any landline data service. Although the mobile software must be slightly modified to perform two-stage dialing, the user is no longer limited to communicating solely with services that happen to have compatible cellular modems (Figure 5.2).

Best known is Microcom Networking Protocol® (MNP), developed by Microcom (based in Norwood, Massachusetts). MNP started out as a proprietary error control protocol, quickly becoming a de facto standard for dial-up modems, and was eventually incorporated (in part) in the CCITT V.42 international standard. MNP Class 3 defines an error-control protocol for asynchronous modems. Start and stop bits are stripped off asynchronous characters (user data) and the data are sent in packets containing an error checking sequence. At the other end of the link, the receiver verifies the data by performing the same error checking calculation and comparing its answer with the one received. If it is different, the receiver requests retransmission of the erroneous packet, which is numbered so the transmitter does not need to wait for each packet to be acknowledged. MNP Class 4 adds adaptive packet sizing, and MNP Classes 5 and 7 add data compression techniques.

While these lower levels of MNP help eliminate data errors and enhance throughput, they do little or nothing to improve the chances of surviving the initial handshake—the most fragile moment during a modem call. MNP Class 10 does, however, and includes several Adverse Channel Enhancements designed specifically for cellular and other hostile links. Negotiated Speed Upshift enables the modem to connect at a lower speed and shift to a higher speed later. Robust Auto-reliable allows multiple attempts at negotiating an error-controlled connection. Dynamic Speed Shifting permits the modem to shift up or down in speed during a connection in response to changing link conditions. Aggressive Adaptive Packet Assembly adjusts packet size to optimize performance at the current error rate (starting out with short packets and lengthening them while error rates are low). And Dynamic Transmit Level Adjustment varies the signal level to match the current cellular link.

MNP 10 faces fierce competition from AT&T Paradyne's Enhanced Throughput Cellular® (ETC) protocol. ETC offers many of the same features as MNP 10, but AT&T has figured out how to get some of them to work even when connected to a non-ETC modem. AT&T claims ETC runs 9,600 to 14,400 without data

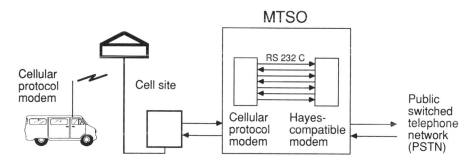

Figure 5.2 Modem pools installed in cellular MTSOs allow mobile users with cellular protocol modems to access virtually any landline data service.

compression, and up to four times faster with it. Other protocols, such as Cellular Data Link Control (CDLC) developed by Millidyne and used primarily in the United Kingdom, may employ FEC, bit interleaving, and other techniques to enhance performance over adverse cellular channels.

Firms such as Air Communications of Sunnyvale, California, have developed integrated voice/fax/data cell phones that optimize both the modem and the radio for data transmission. Air Communications claims its units can connect at speeds of up to 14.4 kbps, with end user throughputs (after compression is applied) of up to 57.6 kbps (Figure 5.3).

In the absence of protocols such as MNP Class 10 or ETC, data calls over analog cellular are significantly less reliable than over dial-up lines. Ordinary modems will fail to connect as much as 50% of the time over cellular channels; cellular protocol modems increase the chances to better than 90%. Given the high cost of cellular air time and the fact that most cellular networks begin charging from the time the user initiates a call, data over cellular can prove a rather expensive proposition. In general, circuit-switched cellular is not well suited to short, frequent transactions, because call setup often takes longer than the data exchange, and the user is usually billed for the first full minute, even if the transaction is completed in seconds.

UPS's package tracking system represents a unique application of circuit-switched cellular. UPS negotiated a deal with four cellular carriers—McCaw Cellular Communications, GTE Mobile Communications, PacTel Cellular, and Southwestern

Figure 5.3 Air Communications' Air Communicator is an integrated voice/fax/data cell phone.

Bell Mobile Systems—who provide unified billing and network management for a total of 49 carriers covering 722 cellular markets across the United States. Using modems supplied by Motorola's UDS division, 60,000 trucks were brought on line in early 1993. Upon delivery of a package, the recipient signs a stylus-input pad, specially designed for electronic signature capture, on the delivery information acquisition device (DIAD). When the driver returns to the cab, he places the DIAD in the DIAD vehicle adapter (DVA); information is exchanged over an infrared link. The cellular telephone modem (CTM) automatically transmits the information to the cellular network; UPS's packet switching nodes are collocated in MTSOs and relay the data to the firm's computer center in Mahwah, New Jersey. Key to the success of this application is the fact that UPS secured special pricing for a high volume of short-duration data calls.

Other users have embraced data over analog cellular because it is compatible with existing applications (or is more readily adapted), can be cost-effectively implemented even on a small scale, and because they feel confident cellular voice service will be around for a long time. Although no one knows the precise number of cellular fax and data users, it is clear that analog cellular is the single most popular mobile data service.

5.4 DATA OVER DIGITAL CELLULAR

Ironically, data cannot be sent over the U.S.'s first-generation TDMA digital cellular networks, as explained in Chapter 4. However, circuit-switched and packet-switched data services are being crafted for both D-AMPS and Qualcomm's CDMA in the United States, and have already been defined for GSM in Europe. Although data over digital cellular will be a late entrant in the U.S. mobile data market, it could nevertheless prove to be a very powerful solution.

TDMA will handle data by splitting conventional modem functions between the mobile subscriber's cell phone and the network. The new Radio Link Protocol 1 (RLP 1) will enable users to send data instead of digitized voice. RLP 1 will also accept standard (dial-up modem) AT commands, which will be acted on either locally (in the subscriber's fax/data cell phone), remotely (at the MTSO), or at both sides simultaneously, as required by specific commands. It will be possible to send data over half-rate, full-rate, double-rate, or triple-rate digital traffic channels.

RLP 1 uses a combination of automatic repeat query (ARQ), which resends data received in error, and FEC, which sends redundant bits to help the receiver repair occasional errors. User throughput on full-rate channels will be 9,600 bps, and triple-rate channels will run 28.8 kbps. Employing CCITT V.42 data compression, users can expect plain text to run 19.2 kbps on a full-rate traffic channel and 57.6 on a triple-rate channel. D-AMPS data services will also include built-in data encryption.

TDMA offers additional benefits to the data user. Unlike most packet radio networks, TDMA will be able to handle high-speed file transfers. It will also be able to transmit error-free faxes at about two pages per minute. And it will interoperate with analog modems and fax machines through special gateways installed in the MTSOs.

N-CDMA will also support data. Qualcomm claims effective user data rates of 6,800 to 7,600 bps have already been demonstrated in field tests. The firm hopes to make support for the Mobile Internet Protocol part of CDMA data standards. A disadvantage of the popular Transmission Control Protocol/Internet Protocol (TCP/IP) is that it is too "chatty" for the radio environment—many packets are sent back and forth even when a small amount of user data is being exchanged. The Mobile Internet Protocol, which requires the use of a special mobile support router, is being developed to make TCP/IP more radio-friendly. As with TDMA, we can expect voice/data phones for CDMA to cost only slightly more than voice-only phones—potentially a big advantage for data over digital cellular compared to the highly touted cellular digital packet data (CDPD) solution, discussed below.

5.5 PACKET RADIO NETWORKS

Like so much wireless technology, packet radio was pioneered by amateur (ham) radio operators. The advantages of packet radio are (1) it provides a protocol for detecting and correcting errors, and (2) it enables multiple users to share the same radio channel. Service providers like packet radio because they can theoretically serve hundreds or even thousands of users with just a handful of channels. End users are attracted to packet radio because they are told it is the most cost-effective solution. But the mobile data industry may have made a strategic blunder when it assumed shared-use, packet radio channels are the best solution to end user needs.

Packet data may either be sent in *datagrams* or over *virtual circuits*. Each datagram contains the recipient's (as well as the sender's) address and is routed independently. In contrast, a virtual circuit entails setting up a specific path over which each packet is relayed for the duration of the session. In either case, the communications links are shared with other users. There is no call setup process for datagrams and very little call setup delay for virtual circuits. Packet switching, therefore, is more efficient than circuit switching (e.g., cellular voice channels) for short transactions. But it is significantly less efficient for handling large volumes of data, because links are shared with other users, and delays may vary.

A random-access protocol is used to manage contention for the inbound (mobile-to-base station) packet radio channel. Certain times (*free periods*) are designated during which any mobile station may begin transmitting. But there is always a chance that two or more mobiles will do so at the same time, resulting in a collision. When this occurs, at least one of the mobile stations must wait until the next free period before attempting another transmission. (Once a mobile gets through to the base station, it may then be assigned reserved time slots during which it can complete its data transmission.) Random-access protocols are highly efficient for serving a large population of users among whom only a few are likely to be active at any given time.

Most of today's packet radio systems use a collision avoidance type of random-access protocol. With an ordinary random-access protocol, when the base station signals the beginning of the next free period, all of the waiting mobile stations will predictably begin transmitting at once. With a collision avoidance protocol, each mobile must select one of several free slots at random for its first transmission. In this way, the first transmissions of several mobile stations are randomly distributed among a group of free slots, ensuring several stations will get through on their first try (Figure 5.4).

Many experts believe that packet radio is the best solution for the bulk of wireless data applications. This is certainly the case for applications like CAD, credit card verification, and personal text messaging, and most of today's wireless data networks are based on packet radio. But it is not obvious that the bulk of wireless data applications in the future will entail short transactions. As notebook and hand-held computers become more powerful (greater processing speed and memory size), the demand for high-speed wireless data networks will grow. Shared channel packet radio networks may not be the best solution for users exchanging lengthy messages containing a mixture of text, digitized sound bytes, graphics, and full-motion video.

(Some will counter that such applications are simply not suitable for mobile data. Others will recognize this as an opportunity for higher performance mobile data networks. One thing we can all agree on: users are not flocking to today's low-speed packet radio services.)

In general, today's packet radio networks offer low throughput and high latency. While packet radio channels typically run at speeds from 4,800 to 19,200 bps, up to 60% of raw throughput is consumed by protocol overhead. (FEC, for exam-

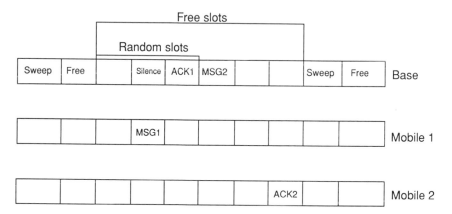

Figure 5.4 Packet radio networks such as Mobitex use collision avoidance protocols. Mobile #1 succeeds in gaining access, while Mobile #2 receives and acknowledges a packet.

ple, typically reduces end user throughput by 30%.) The maximum instantaneous end user throughput over a packet radio channel, therefore, is usually about 50% of the link rate. On a moderately loaded channel, the average end user throughput is usually less than 20% of the link rate. And the average round-trip delay is usually in excess of several seconds, which is the main reason why most landline data communication applications will not work over packet radio networks; they expect response times of, at most, a few hundred milliseconds.

But there are other reasons why landline data communication applications do not work over wireless networks. Mobile users may appear and disappear on the network as they move in and out of radio coverage. When the user is in a fringe coverage area, the quality of the communications link may fluctuate wildly. But even when the user is well within range, most packet radio networks simply do not offer the throughput or response time required for applications involving interactive screen editing or large file transfers.

Unfortunately, today's packet radio network operators and equipment manufacturers must spend a substantial amount of their time encouraging and assisting the development of mobile-enabled applications. While several firms have stepped forward with software tools intended to streamline development, average corporate customers still spend considerable time and money customizing their applications. Nevertheless, packet radio is the technology of choice among the three major U.S. players in nationwide mobile data: CDPD, ARDIS, and RAM Mobile Data. Packet radio technology is also being used to carry data over analog SMR networks. These four solutions are compared in Table 5.1. Let us take a closer look at each.

Table 5.1
Comparison of Mobile Data Services Provided by Cellular Digital Packet Data, ARDIS, RAM Mobile Data, and Racotek

	RAM Mobile Data	*ARDIS*	*CDPD*	*Analog SMR*
Major supporters	BellSouth, Ericsson-GE	Motorola	McCaw Cellular, IBM	Racotek, Motorola
Link rate/ user rate	8,000 bps/ <2,000 bps	4,000 bps/ <1,000 bps	19,200 bps/ <5,000 bps	4,800 bps/ <2,000 bps
Strengths	Coverage, 3rd party support	Inbuilding coverage	Leverages cellular infrastructure	Leverages SMR infrastructure
Weaknesses	Latency, arcane protocol	Low speed, arcane protocol	Must borrow cellular spectrum	Metropolitan area solution
Status as of late 1994	Nationwide service, 12,000 + subscribers	Nationwide service, 35,000 + subscribers	Limited availability	5,000+ subscribers

5.6 CELLULAR DIGITAL PACKET DATA

For several years, the U.S. cellular telephone industry expressed interest in the mobile data market but refused to do anything, saying it would not consider any technology that impinged on its bread-and-butter voice business. Then IBM stepped forward, with strong support from McCaw Cellular Communications, with a solution it claimed would have no impact on cellular's voice traffic capacity, and would entail no replacement of existing hardware or software. IBM dubbed its solution CelluPlaN II.

The CDPD Group was formed in 1992 to develop an open specification based on IBM's concept, and the CelluPlaN II moniker was dropped. This was the most important announcement in the history of mobile data—data was finally recognized as a key growth area for the world's largest cellular telephone market. The original CDPD Group was composed of IBM, nine leading cellular carriers, and retail giant Sears, Roebuck and Co.

The key to CDPD's acceptance was a technique called *channel hopping*. The idea was simple: packet data could be supported within cellular telephone's existing spectrum allocation, leveraging its existing infrastructure, by using idle voice channels. In other words, CDPD would exploit trunking inefficiencies. On any given channel, the lag between the termination of one voice call and the initiation of the next creates a time period during which the channel may be momentarily seized for packet data transmissions—even if voice users are always given priority. CDPD would rely on "RF sniffers" to monitor each channel within a cell; when a channel became idle, it could be borrowed for data, but it would be relinquished immediately when a new voice user appeared. In this way, CDPD service would hop from one channel to another.

(According to IBM, periods during which all of the channels within a cell are busy handling voice users are extremely rare and short-lived. Remember, voice users appear and disappear not only at the initiation and termination of each call, but as a result of cell handoffs.)

CDPD employs Gaussian minimum-shift keying (GMSK) modulation to achieve a radio link speed of 19,200 bps. Instantaneous end user throughput is 55% to 60% of the link speed, or about 11,000 bps. Due to channel contention, however, the average user throughput will be much lower, perhaps 2,400 bps. CDPD is designed to support cell handoffs, intersystem handoffs, and automatic nationwide roaming. Both connection-oriented (virtual circuits) and connectionless (datagram) services will be offered. Connection-oriented services include both reliable (network provides error control) and nonreliable (application provides error control) modes.

A draft specification was made available to the industry for comment and review shortly after the CDPD Group was announced. Version 0.8 of the open specification was released on 22 March 1993—less than one year later. The final specification was published in July 1993 with the hope that it would become an industry standard. But while progress towards a final specification was rapid, implementation has been much slower. By September 1994, only one cellular carrier had announced limited commercial availability of CDPD service (Bell Atlantic Mobile Systems).

CDPD has clearly run into problems. The largest individual mobile data customers—companies like Federal Express and UPS—tend to have nationwide requirements. But in the United States, cellular carriers are licensed on a regional basis. Since not all carriers have exhibited the same level of commitment, there is much confusion regarding how quickly CDPD will be deployed nationwide. It is also not clear how CDPD will be marketed, supported, and priced. (No doubt, we will eventually see intercarrier agreements for CDPD, just as there are intercarrier agreements for voice service.)

Channel hopping has become embroiled in controversy. Some of the carriers have conducted tests using dedicated channels. For carriers, the use of dedicated channels is less complex. For end users, dedicated channels reduce the chances of lengthy transmission delays and will probably enable lower cost radio modems. The obvious disadvantage of using dedicated channels, however, is that CDPD might impact voice capacity after all. Most carriers now say they plan to use both hopping and dedicated channels, depending upon local network and customer requirements.

The CDPD Group claims the construction of a nationwide network will be relatively quick and inexpensive, because it will use existing cell sites, antenna towers, and dedicated phone lines back to the MTSO. Perhaps an even bigger advantage of CDPD over data-only nationwide networks, however, is that mobile data is merely incremental revenues for cellular carriers. Revenues derived from voice services will likely subsidize data during its first critical years. In contrast, data-only networks operated by ARDIS and RAM Mobile Data are under intense pressure to achieve short-term results, because data accounts for all of their operating revenues and costs.

Will CDPD become the dominant mobile data platform in the United States? While there is considerable support for CDPD, it has not been as widespread as had been expected. Dial-up modem manufacturers, PC communications software developers, and others appear to be taking a wait-and-see approach regarding CDPD product development. McCaw Cellular Communications has been the most outspoken carrier, but has failed to meet the aggressive deployment schedule it had promised. Critics (as well as competitors) have had a field day accusing CDPD players of hype (i.e., overselling their solution).

It seems unlikely that the growth of the mobile data market has been retarded by the exaggerated claims of a few CDPD proponents. If that were the case, the entire computer and communications markets would have collapsed long ago! More likely, the source of ARDIS's and RAM Mobile Data's trouble is closer to home. Customers complain that adapting their applications to ARDIS's and RAM Mobile Data's services, with their arcane protocols, often entails a long and arduous software development effort. Because it is closely based on landline standards such as TCP/IP and offers somewhat higher throughput, it may be significantly easier to get applications up and running over CDPD.

5.7 ARDIS

ARDIS (Advanced Radio Data Information Service) began as a joint venture between IBM and Motorola. The network was originally built by Motorola in 1984 for IBM's National Service Division, and was designed to provide optimal in-building coverage. IBM's original applications included CAD, parts ordering and tracking, service contract entitlement checking, invoicing, and general messaging. But in 1994, IBM sold its 50% interest in ARDIS to Motorola.

ARDIS introduced its two-way wireless data service in over 400 metropolitan areas across the United States in 1990. Since opening its door to non-IBM customers, ARDIS has had its most success with other field service organizations. Subscriber growth has been slow, however, rising from the initial 22,000 IBM users to slightly more than 35,000 users in about four years. The firm has been criticized for focusing exclusively on vertical (i.e., industry-specific) applications and neglecting to build horizontal (i.e., cutting across multiple industries) markets. But it is clear that most of today's mobile data business is found in vertical markets.

ARDIS offers excellent building penetration thanks to the use of single-frequency overlapping radio cells (Figure 5.5). The use of overlapping cells increases the probability that a portable user's transmissions will get through to a base station from within a building. But it also reduces the network's overall capacity, because base station transmitters must be frequently turned off (for 0.5 to 1 sec) to prevent

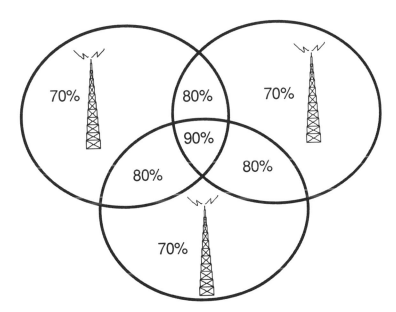

Figure 5.5 ARDIS's overlapping radio cells increase the probability of reaching a base station from within a building.

them from interfering with each other. As a result, ARDIS has been called a "two-way paging network" by critics, and recommends that customers refrain from attempting to send files containing more than 10,000 bytes of data.

Each ARDIS base station transmits with 40W of power and covers the area within a 15- to 20-mile radius. ARDIS's MDC 4800 radio frequency links run 4,800 bps over 25-kHz-wide duplex channels (45 MHz apart) in the 800-MHz band. (ARDIS's 25-kHz channels give it an advantage over RAM Mobile Data, which uses 12.5-kHz channels.) Portable units transmit with up to 4W of power.

ARDIS claims its network, which contains over 1,300 base stations, encompasses over 90% of U.S. business activity. Each base station is tied into a radio network controller, which in turn is tied into the network control center (NCC) located in Lexington, Kentucky. The NCC controls all message routing (all messages travel all the way up the hierarchy), network management, and billing and accounting. The customer's computer center is linked to the ARDIS NCC via dedicated asynchronous (mainly used for pilot projects), bisynchronous, X.25, or Systems Network Architecture (SNA) lines.

Motorola introduced its InfoTAC radio modem/personal communicator for ARDIS in September of 1992 (Figure 5.6). The 18-ounce InfoTAC offers two interface modes: transparent mode (responds to dial-up modem AT commands) and native mode (a more powerful packet radio-specific mode). Software can be downloaded into InfoTAC's flash read-only memory (ROM) from an external PC to customize responses, network protocol, and channel scan lists. The InfoTAC is a

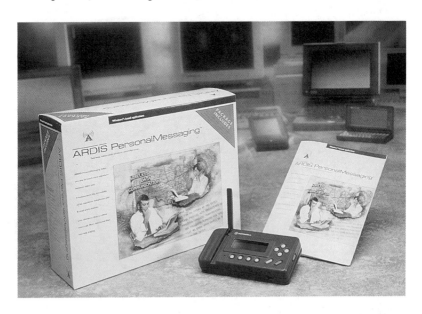

Figure 5.6 ARDIS's *PersonalMessaging* package enables peer-to-peer workgroup messaging.

protocol-agile radio modem for use over ARDIS (U.S.), Bell ARDIS (Canada), Hutchison Mobile Data (Hong Kong and U.K.), Modacom (Germany), and various Mobitex networks.

ARDIS's service costs are based on the size and volume of packets used by each subscriber. Packets range from about $0.08 to $0.17 each, with the maximum-sized packet containing 240 characters. As can be seen immediately, data transmissions over ARDIS are much more expensive than over landlines, and users have plenty of incentive to tune their applications to require as little data flow as possible.

ARDIS plans to upgrade most of its base stations to 19,200 bps using Motorola's RD-LAP protocol and DataTAC base stations. RD-LAP employs four-level FM signaling (as opposed to MDC-4800's two-level FM), and supports packets of up to 512 bytes long—twice the MDC-4800's 256 bytes. However, ARDIS has indicated that the upgrade will primarily increase network capacity rather than enhance throughput for individual users.

In an attempt to counter Mobitex networks, ARDIS also announced the formation of a Worldwide Data Network Operators Group consisting of ARDIS, Deutsche Bundespost Telekom, Hutchison Mobile Data (U.K.), Bell-Ardis, and Hutchison Mobile Data (Hong Kong). Bell-Ardis in Canada is a joint venture between BCE Mobile and Motorola Canada and covers the 26 largest metropolitan areas in Canada.

ARDIS's (as well as RAM Mobile Data's) small number of subscribers should be a cause of concern for the entire mobile data industry. Some analysts have commented that ARDIS's and RAM Mobile Data's services are too costly. Both firms, however, have experimented with introductory and special package pricing with little success. ARDIS has been criticized for a lack of network capacity, but what it suffers from is a lack of subscribers. A new generation of personal communicators, leveraging new types of messaging services, could bring about a dramatic change in prospects for ARDIS.

5.8 RAM MOBILE DATA

ARDIS's major competitor, RAM Mobile Data, is backed by BellSouth and operates a nationwide mobile data network based on the international Mobitex specification. RAM has performed yeoman's work in developing horizontal markets for wireless electronic mail and messaging, partnering with companies such as Intel (marketing a radio modem through its retail channels), Lotus Development Corporation (e-mail software), AT&T EasyLink (e-mail services), RadioMail Corporation (wireless-to-wired gateway services), Simware (wireless IBM terminal emulation software), and Oracle Corporation (wireless access to corporate databases). RAM's Wireless Messaging Business Unit was formed in 1992 to foster the development of "shrink-wrapped solutions" for wireless e-mail.

Mobitex is an open specification for two-way land mobile communications (voice and data) conceived by Swedish Telecom (with support from Swedish radio giant L. M. Ericsson), but implemented as a data-only service in the United States and Canada (Rogers Cantel). The first Mobitex network was built in Sweden in 1986; other networks have been deployed in Norway, Finland, the United Kingdom, and France. The Mobitex Operators Association plans to support international roaming—a feature that will be hard to provide due to different frequency allocations in different countries. Mobitex combines trunking, frequency reuse, and packet radio to create a high-capacity network for short- to medium-sized data transactions. While RAM's total capacity compares favorably to ARDIS's, it suffers similarly from low throughput and high latency, and the firm discourages file transfers in excess of 20,000 bytes.

RAM's network was constructed under a waiver of the FCC SMR Rules and now boasts over 800 base stations in the top 100 MSAs in the United States. While ARDIS enjoyed an early lead in geographical coverage and building penetration, RAM has focused on providing excellent coverage in airports, convention centers, and hotels in major cities. RAM's 12.5-kHz-wide channels are located in the 896- to 901-MHz (mobile transmit) and 935- to 940-MHz (mobile receive) bands, with each transmit/receive pair separated by 39 MHz. The radio link runs 8,000 bps.

Mobitex is a hierarchical network consisting of a network control center, main exchanges, area exchanges, and base stations. Unlike ARDIS, traffic flows only as high up the hierarchy as necessary; that is, messages between two mobile stations operating within range of the same base station are switched locally. Only activity reports (for billing and user tracking) are sent to the network control center.

Mobitex handles packets containing up to 512 bytes of ASCII text or binary data. Over 40% of the RF link rate is dedicated to protocol overhead—mostly FEC. RAM claims messages under 100 bytes in length are typically delivered within 2 to 4 sec. Billing is on a monthly basis and consists of a fixed subscription plus per-packet charges. Packet charges start at $0.03.

Although Mobitex is based on an open specification, obtainable by any potential manufacturer, Ericsson-GE Mobile Data is by far the largest supplier. The mobile or portable terminal accesses the network by registering on the closest (strongest receive signal) base station. The first trunked channel handles both session setup/teardown and data traffic. Additional channels are activated as traffic demands. Ericsson-GE Mobile Data's Mobidem M1090 portable radio modem offers an RS-232 interface for laptop, notebook, and palmtop computers. The Viking Express is a small vinyl pouch that contains an Ericsson Mobidem linked to an HP 95LX palmtop running RadioMail Remote software, which sells for under $1,000.

Mobitex supports nationwide roaming (a feature that was added to ARDIS's network in 1993). Each base station broadcasts a list of frequencies of nearby base stations. During a base station SWEEP cycle, each of the base stations transmits in sequence, allowing mobile stations to measure and compare their signal strength. If a mobile determines over two SWEEP cycles that another base station has a stronger signal, the mobile registers onto the new base station. The Mobitex terminal specification also requires that a mobile terminal notify the network just before it is powered down; terminals are designed to send an INACTIVE message immediately after the power switch is turned off.

Mobitex's 24-bit address field supports up to 16 million addresses. Terminal, closed user group (CUG), and personal subscriptions are supported. A single terminal subscription can support multiple input/output devices such as printers and bar code readers. CUGs allow end user organizations to operate as if on their own private radio network.

Like ARDIS, RAM Mobile Data has so far had modest success, claiming only 12,000 subscribers as of September 1994. There is little doubt, however, that RAM's missionary work in wireless e-mail and messaging will pay off. The question is, however, how long the firm's owners will be willing to nurture the market. While ARDIS has been pressured to address horizontal markets, RAM has come to recognize the importance of vertical markets as a source of revenue during the horizontal markets' gestation period.

5.9 DATA OVER SPECIALIZED MOBILE RADIO

Another interesting player is Minneapolis-based Racotek (an acronym for "radio computer technology"), who believes the early applications for mobile data will be predominantly oriented toward vertical markets, and that customers want solutions that are independent of any particular wireless infrastructure.

Racotek provides a transaction-oriented service called RacoNET that overlays existing specialized mobile radio systems. Since relatively little airtime is needed to handle data transactions, RacoNET offers SMR operators an opportunity to increase revenues even on systems that are already operating near full voice capacity. RacoNET service is also unique because it is billed by Racotek; the firm dials into its installed systems to retrieve usage data, paying the SMR operators commissions on their data traffic.

Racotek's CEO Richard Cortese believes mobile data will succeed much like the microcomputer: first penetrating vertical markets that are willing to invest more to become the early beneficiaries of an emerging technology. Racotek is working with value-added resellers (VAR), systems integrators (SI), and independent software vendors (ISV) to create or modify applications to work under Racotek's Mobile Network Operating System (R/MNOS). Cortese believes mobile data customers are being asked to divine which mobile data network is best, when what they are interested in is achieving the right level of performance at the right cost. Racotek has developed powerful simulation tools to help customers identify the best radio infrastructure for their required geographical coverage and traffic characteristics. This approach contrasts with ARDIS and RAM Mobile Data, each of whom is trying to convince customers to standardize on a specific infrastructure.

Racotek claims trunked channels in over 1,100 locations across the United

States are compatible with its service. The firm has formed strategic alliances with Motorola, EF Johnson, and other SMR players to market its service and products. Racotek offers several different billing options, with flat rates available for various volume ranges (e.g., up to 225,000 characters for $50 per month). Racotek becomes the customer's point of contact for both voice and data service, with multiyear contracts available to limit future price increases.

R/MNOS is an open-system client server architecture designed for portability across mobile devices, host platforms, and radio infrastructures—what Racotek's vice president of engineering Isaac Schpancer calls "Novell over the airwaves." The RacoNET hardware/software solution consists of a Communications Gateway, Mobile Communications Controller (MCC), and R/MNOS software. The Communications Gateway can be configured for a variety of host platforms and acts as a wireless server. The MCC can receive and store up to 2.3 MB of data. The built-in radio modem runs 4,800 bps using phase-shift keying (PSK) and includes automatic error correction.

Racotek is trying to give customers what they want: turnkey mobile data solutions. The main challenge for Racotek will be to ensure its approach is embraced by Nextel, who is building a nationwide ESMR network that will ultimately offer both circuit-switched and packet-switched digital data services. Like the early microcomputer industry upon which Racotek models itself, mobile data is sure to undergo fantastic changes over the next decade.

5.10 MOBILE DATA IN EUROPE

Thanks to Mobitex networks in Scandinavia, Europe has more years of experience in marketing mobile data services than the United States. But with the spin-off of ARDIS as a service and the construction of RAM Mobile Data's network, the United States has surged way ahead in terms of geographical and population coverage. In fact, there are currently more mobile data users in the United States than the rest of the world combined.

Mobitex has a foothold in Norway, Finland, Sweden, the United Kingdom, Netherlands, and France. Motorola's DataTAC architecture—the newest version of the architecture employed by ARDIS in the U.S—is supported by the Deutsche Bundespost Telekom's Modacom in Germany and Hutchison in the United Kingdom. Notably, Modacom is continuing to expand its network to cover 80% of Germany, despite the fact that Germany has been more aggressive than any other country in deploying GSM digital cellular, which promises a full suite of data services. The ETSI is also working on another grand project: the trans-European trunked radio (TETRA) standard for two-way packet-switched data, circuit-switched data, and circuit-switched voice. This is roughly equivalent to ESMR in the United States.

As in the United States, it is believed that Europe's mobile data market will become large enough to support multiple service providers. But the two markets present an important test of different philosophies regarding the development and commercialization of new technologies. Particularly at issue are the role of government megaprojects and committee-based standards. While few would doubt that the free market is the best context for the development and commercialization of consumer products, many believe the telecommunications infrastructure requires more planning and the establishment of early standards. We should know, within the next decade, which approach is correct.

5.11 HOW MOBILE DATA WILL SUCCEED

Most of today's mobile data networks are essentially wireless Telex services. Unfortunately, they are designed to meet yesterday's requirements—only on a larger scale. As computer and data communications users have proven over and over, the more popular a service or product becomes, the more users demand of it. Network operators have fooled themselves into believing mobile data defines a collection of low-speed applications simply because that is how it was in the past.

From microcomputers to cell phones, new products tend to penetrate vertical business markets first, horizontal business markets second, and consumer markets last. While attempts to leapfrog the vertical market phase are commendable, the amount of development effort needed to establish broad horizontal markets is equal to or greater than the customer hand-holding and nurturing required in the vertical applications. There is no reason, however, why flexible platforms designed for horizontal markets cannot be adapted to the urgent needs of today's vertical markets.

Today's mobile data networks are plagued by low-throughput, expensive radio modems, high per-packet charges, power-consuming portable units, and arcane packet radio protocols. The mobile data market awaits a breakthrough in network performance and cost-effectiveness. The market will likely bifurcate into high-speed access to enterprise networks and low-speed two-way messaging. When mobile data networks can provide the throughput of medium- to high-speed dial-up modems at low monthly flat rates and work with off-the-shelf or slightly enhanced dial-up communications software, the market will surely explode.

SELECT BIBLIOGRAPHY

[1]   Arnbak, J. C., "The European (R)evolution of Wireless Digital Networks," *IEEE Communications*, September 1993.

[2]   Brodsky, I., "Technologies for Mobile Datacomm," *Business Communications Review*, February 1992.

[3]   Brodsky, I., "Wireless E-Mail: A 'Killer' Application for Pen Computers?" *Pen Magazine*, February 1992.

[4]   Brodsky, I., "Wireless MANs/WANs Offer 'Data To Go,'" *Business Communications Review*, February 1991.

[5]   Brodsky, I., "Dialing Data Via Mobile Telephones," *Data Communications*, October 1989.

[6]   Brodsky, I., "Wireless Data Networks and the Mobile Workforce," *Telecommunications,* December 1990.

[7]   Cortese, R., CEO of Racotek, interview on 21 May 1993.

[8]   Datacomm Research Company, *Portable Computers and Wireless Communications*, industry research report, Wilmette, IL, 1993.

CHAPTER 6

▼▼▼

PERSONAL
COMMUNICATIONS SERVICES

6.1 CELLULAR FOR THE REST OF US?

In 1979, the FCC took a major gamble and reallocated the upper 14 UHF-TV channels to land mobile radio. Spectrum was commandeered from the powerful television broadcast industry in order to establish the cellular telephone and specialized mobile radio (SMR) services. In a gush of enthusiasm, AT&T predicted there would be more than 900,000 cellular telephone users by the year 2000.

But AT&T was wrong. Cellular telephone turned out to be a modern-day gold rush. By early 1994, there were already more than 15 million cellular telephone subscribers. To many, it appeared the government had created—through its wisdom and foresight—an entire new industry.

Now, telecommunications regulators throughout the world are falling over each other in a mad dash to launch the next mobile communications industry: personal communications. In the United Kingdom, government officials hope to catapult domestic high-technology players to the forefront of this strategic global market. In the United States, the Clinton administration anticipates PCS will generate 300,000 new jobs and help offset the federal budget deficit by raising $10 billion through spectrum auctions.

There is just one problem: no one is quite sure what personal communications is. When the FCC allocated 160 MHz to the new PCS in late 1993, it defined PCS as ". . . a variety of new mobile services, technologies, and equipment that will operate at home, at work or on the street." In other words, they do not know either.

PCS is different things to different people. To LECs, PCS is another way to access the public telephone network. To regional cellular telephone carriers, PCS could be a path to nationwide coverage, or perhaps just a new packaging and billing option. To outsiders, PCS is an opportunity to invade previously closed markets— busting up long-standing LEC monopolies and cellular duopolies.

The personal communications market was first suggested when business analysts noticed a wide gap in price/performance ratio between the residential cordless phone and cellular telephone (Figure 6.1). Cordless phones are found in well over 25% of U.S. households, while cellular has only managed to penetrate about 8% of the U.S. market. According to analysts, there is probably a market for a product that costs less than cellular but outperforms residential cordless—cellular for "the rest of us."

Others believe PCS is destined to become more than just cheaper, better cellular. PCS will allow us to call people, not things. To paraphrase Arthur C. Clarke, the time is coming when you will call someone, and if they do not answer, you will know they must be dead. Personal communications requires a new type of intelligent network—a network smart enough to recognize individual users, learn their preferences, and serve them wherever they appear. But there is also the fear that mobile communication networks will leave us nowhere to hide.

Consumers associate PCS with the palm-sized communicators used in the *Star Trek* TV series. These communicators display impressive range, never seem to require fresh batteries, and can establish an instant link to whomever the user calls out. But according to the cellular telephone industry, PCS is already here: handheld

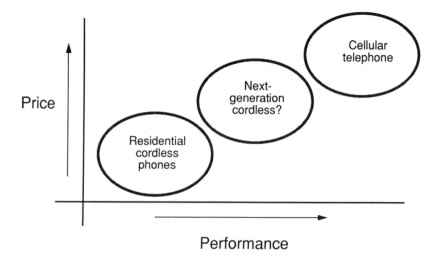

Figure 6.1 The price/performance gap between cellular and residential cordless telephones.

phones dominate new sales. There are also consumer-oriented service packages available; in some cases, cellular service is cheaper than comparable landline service.

For U.S. long-distance carriers, PCS could become an "end run" around LECs and local access charges. AT&T acquired McCaw Cellular Communications—all the while denying the deal was an attempt to reenter the local telephone business. MCI tried to buy a chunk of Nextel, an ESMR operator that says it is in the personal communications business, but the deal fell apart. The U.S.'s third major long-distance carrier, Sprint, also owns a major cellular operator and will no doubt become involved in PCS.

Another PCS goal is a single handset that operates in the home, office, and public areas. A *common air interface* (CAI) that cuts across public and private systems will help differentiate PCS from cellular. When users are at home or in the office, their handset will automatically default to a private system (no airtime charges); but when they are beyond the range of a private system, the handset will register on a public system.

In Britain, second-generation cordless telephones (CT2) were expected to serve as wireless callboxes (pay phones). The inconvenience of locating and using coin-based pay phones—often targets of vandals—suggested pocket phones would enjoy immediate success. While Telepoint (as the service became known) flopped in Britain, it fared somewhat better in Hong Kong. In the United States, wireless pay phone service has been proposed for truck stops along interstate highways.

PCS may become the first two-way multimedia wireless service. Using digital radio technology, broadband PCS has sufficient bandwidth for voice, text, fax, graphics, and video transmission. PCS could be used by pocket information appliances to send and receive faxes, pull electronic newspapers and magazines out of the air, and send and receive graphics files. PCS players will have a hard time competing with cellular operators for voice business, but they should be able to construct networks uniquely capable of delivering wireless multimedia services.

But none of these descriptions captures the true essence of personal communications. What we are talking about is a telecommunications facility we can reach for anytime, anywhere; perhaps it would be more accurately called *personalized* communications. It is not a single entity, but a constellation of services including paging, residential cordless telephony, data broadcasting, cellular telephony, and mobile data. The narrowband (930 MHz) and broadband (1.8 GHz) PCSs are simply the latest additions to a growing list of wireless services designed to empower individuals. The personal communications revolution does not hinge on the success of PCS. But with the help of these new services, it will triumph sooner.

6.2 THE EVOLUTION OF PCS

The United Kingdom deserves much of the credit for developing the concept of next-generation cordless telephony. Noting the performance limitations of analog cordless phones, and the high cost of cellular service, several firms began meeting under the sponsorship of Britain's Department of Trade and Industry (DTI) in 1984 to discuss a possible second-generation technology. They hoped that by hammering out technical standards and assigning frequencies early, Britain would secure the leading role in what was certain to become a vital global market. But as the idea caught on, several competing varieties of cordless technology emerged.

First-generation cordless telephones (CT1) employ analog transmission and typically operate in the high-frequency (HF) band. Although CT1 phones are susceptible to noise and interference, have limited range, and offer little if any privacy, a consumer mass market has arisen due to their low cost.

In the United States, analog cordless phones operate in the 49-MHz band under FCC Rules Part 15. Although extremely popular in the home, cordless phones have enjoyed little success in the office. There are many technical explanations for their lack of acceptance among business users. CT1 phones do not have the capacity for high-density environments, lack roaming capability, and offer no privacy.

A digital cordless telephone (DCT), it was theorized, would solve these problems. Digital also promises immunity to interference and noise, and support for data. If the United States could become the leader in cellular telephone by being first, the United Kingdom could become the leader in digital cordless by likewise being first.

6.2.1 CT2

The British defined CT2 around FDMA/TDD technology. FDMA divides the radio band into distinct channels, one for each conversation. TDD divides each channel into alternating talk and listen time slots for duplex operation. Critics pointed out, however, that FDMA/TDD requires separate base station radios for each channel and costly duplexers in the handsets.

Three major applications were envisioned for CT2 phones: (1) residential cordless phones with enhanced performance, (2) wireless key telephone systems (WKTS) and wireless PBXs (WPBX), and (3) pocket phones for outdoor wireless pay phone service—what became known as Telepoint.

In 1987, the U.K.'s DTI allocated forty 100-kHz-wide channels from 864 to 868 MHz for Telepoint, hoping to get a head start against potential competitors from the United States, Japan, and other countries. In the same year, CT2 was codified in two published standards: MPT 1334 and BS 6833. These standards were intended to prevent interference between what were expected to be four dissimilar licensed systems, and they defined how all would interface with the PSTN.

Telepoint provided cordless access to the public telephone network at strategic locations such as banks, train stations, and restaurants. CT2 was designed for one-way calling and did not support cell handoffs. In other words, the user could originate but not answer calls, and had to stay within range of the same base station for the duration of a call. These constraints, however, should not be construed as inherent in FDMA/TDD technology.

Although Telepoint service appears to have succeeded in Hong Kong and has shown signs of life in France, all four of Britain's licensed Telepoint operations have been forced to shut down. Three of the four have deployed proprietary systems; handsets designed to work with the base stations on one system could not be used with base stations from another. Ironically, the three committed to implementing a common air interface (CT2/CAI), detailed in the MPT 1375 standard, just before shutting down. CT2/CAI defines the modulation technique, data protocol, and speech encoding algorithm necessary to ensure interoperability. (The fourth operator, Hutchison Personal Communications, deployed only after the first three operators had shut down, making a CAI irrelevant.)

Attributing CT2's failure to technical limitations, Northern Telecom (who was a participant in one of the original Telepoint consortia) developed an enhanced version of CT2 technology called CT2Plus.

6.2.2 CT2Plus

Perhaps desiring to secure an advantage for Canada, Northern Telecom developed CT2Plus, a technology that overcomes many of the limitations of CT2. CT2Plus Class 1 is backwards-compatible with CT2/CAI, but also supports cell handoffs, two-way calling (inbound as well as outbound), calling-party identification, inrange service indicator, and data transmission. CT2Plus Class 2 adds a common signaling channel for faster call setup. Not surprisingly, CT2Plus has been selected as the DCT standard for Canada, and Northern Telecom has developed a family of CT2Plus-based WPBXs.

But CT2Plus is not without its drawbacks. Once a CT2 or CT2Plus voice channel is established, the handset occupies that channel 100% of the time. It cannot monitor other channels in its own or adjacent cells without interrupting the speech path. To effect a handoff, it must discontinue operation on the original channel and resume operation on a new channel. (As in analog networks, adjacent cells in an FDMA/TDD cellular network cannot normally reuse the same frequencies.)

L. M. Ericsson has developed a third-generation cordless telephone system (CT3) that offers continuous monitoring of other channels and seamless handoffs.

6.2.3 CT3

TDMA/TDD is the underlying technology of CT3. The radio band is divided into frequency channels, and each channel is divided into time slots. A single base station radio handles multiple conversations on a given frequency channel, reducing the cost of cell site equipment.

CT3 technology specifies four 1-MHz-wide radio carriers. Each radio carrier is divided into 16 time slots, a pair of which are required for each duplex voice channel (talk and listen). There are 32 voice channels available per base station. CT3 relies on ultrasmall *picocells* to obtain the capacity required in high-user-density environments, such as offices.

Since a CT3 handset only communicates during 2 of 16 time slots, it is free to handle other tasks during the remaining 14 time slots. (Remember, voice bits are briefly stored and sent in bursts.) Normally, the CT3 handset monitors signal strength on other frequencies during this time. When a base station handoff is required, the CT3 handset temporarily sets up two parallel voice paths, avoiding any need to interrupt the voice channel. And because a single CT3 time slot runs 32 kbps, it is well suited to data applications; by allocating eight time slots, data speeds up to 256 kbps may be achieved.

Ericsson developed CT3 as a proprietary forerunner to the official third-generation Digital European Cordless Technology (DECT) standard, to which the firm is also committed.

6.2.4 DECT

This comprehensive standard addresses residential cordless phones, wireless office telephone systems, Telepoint-like public access systems, and wireless LANs, and is subscribed to by the 18 member countries of the CEPT. DECT has been allocated the region from 1,880 to 1,900 MHz throughout Europe.

DECT specifies 10 channels, each divided into 24 time slots, for 12 two-way voice conversations per channel. Channels need not be preassigned to cell sites. The handset may activate any channel it determines to be free. In this way, DECT systems can quickly adapt to changes in propagation or traffic load. The integrity of signaling data (e.g., for call setup and handoff) is ensured through the use of an ARQ protocol employing a 16-bit cyclic redundancy check (CRC). The voice bit stream is not error-protected, however, because the delays associated with retransmitted voice packets are intolerable to human listeners. Each time slot runs 32 kbps; each channel (including signaling and overhead) operates at a composite speed of 1.152 Mbps.

While DECT targets applications in limited geographical areas, another TDMA system (DCS-1800) is defined for wide-area PCNs serving both mobile and fixed users.

6.2.5 DCS-1800

In the United Kingdom, over 90% of local wireline telephone access is provided by one company: British Telecom. The only other local access provider is Mercury Communications. Although PCNs were first conceived to provide mobile services, their ultimate success in the United Kingdom may have more to do with challenges to British Telecom's near monopoly.

Like Telepoint, PCNs were a child of the U.K.'s DTI, who in 1989 announced it would accept PCN license applications. The DTI stipulated that PCNs must be based on ETSI standards, encouraging interoperability throughout Europe and en-suring economy of scale for manufacturers of subscriber terminals. The licensees proposed GSM (900 MHz) digital cellular technology modified for operation at 1,800 MHz. PCNs are also planned in France, under the regulations of the Direction de Reglementation Generale (DRG), and in Japan, where Personal Handy Phone service based on (non-GSM) TDMA technology has been allocated spectrum in the 1.9-GHz band.

DCS-1800 was developed by the ETSI's GSM technical committee and ap-proved in February 1991. DCS-1800 supports multiple operators, both in terms of offering a choice of carriers in any given location and supporting wide-area roaming. Nationwide roaming is important because one operator cannot be expected to pro-vide ubiquitous coverage. The 1,710- to 1,880-MHz band was allocated by CEPT for use throughout Europe and consists of two 75-MHz-wide bands, 20 MHz apart.

PCN proponents envision small cells, low-power handsets, and base station antennas close to the ground. While GSM was primarily defined for fast-moving vehicle-based users, PCN is mainly intended for pedestrian users. However, even governments and standards committees cannot stop PCN users from communicating while in their automobiles. Therefore, PCNs also address fast-moving users and the frequent handoffs they may require. In addition, with multiple operators and high user densities, PCNs must be designed to tolerate one carrier's user operating at full power next to another carrier's base station, and different subscribers operating within very close range of each other (Figure 6.2).

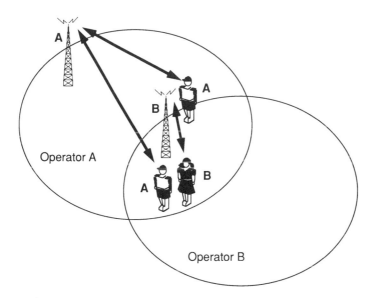

Figure 6.2 In order to coexist, two GSM operators must tolerate operation in close proximity.

Paralleling Europe's push towards PCNs based on GSM-derived technology, AT&T has announced PCS technology for the U.S. market based on the two North American digital cellular standards: IS-54 (TDMA) and IS-95 (CDMA). The move is intended to ensure new wireless products will interoperate with North America's existing wireless and wireline networks. More specifically, it could enable McCaw Cellular to offer nationwide service via dual-frequency (800 MHz/1,800 MHz) handsets.

6.2.6 DS-SS and N-CDMA

In the United States, PCS licensees must (at least initially) coexist with incumbent fixed microwave users. For this reason, many engineers favor spread spectrum technology. Two firms specializing in spread spectrum—Qualcomm and Omnipoint Corporation—have formed strategic alliances with potential PCS players.

Qualcomm claims its N-CDMA technology requires less than half as many cells as a DCS-1800 network providing the same capacity and coverage. Because CDMA channels can be reused in adjacent cells, no frequency planning is required. Qualcomm also claims greater range and building penetration than DCS-1800 using the same antenna and power level. But perhaps Qualcomm's assertion that CDMA requires a lower average capital investment per subscriber at market penetration rates below 20% is the bottom line. (While it is clear PCS could become a "cheaper, better" cellular service once it achieves high market penetration, it is less clear at low market penetration.)

Qualcomm plans to pursue PCS with its core and extended CDMA systems. The core system uses a 14.4-kbps air interface and 13-kbps codecs (voice coders). The extended system uses a 76.8-kbps air interface, works with higher quality codecs at 16 and 32 kbps, and supports feature enhancements such as challenge-response authentication. AT&T, Northern Telecom, and Motorola plan to introduce PCS products based on Qualcomm's CDMA technology in 1995.

Omnipoint Corporation is taking a different approach. Its DS-1900 system uses a combination of TDMA and DS-SS to facilitate spectrum sharing and handset-controlled cell handoffs. DS-1900 handsets will work over both licensed (public access systems) and unlicensed (wireless PBX) systems, and can support the rapid handoffs expected when traveling at high speeds through microcells. DS-1900 handsets will support data rates from 8 kbps (voice) to 500 kbps (data) (Figure 6.3).

Both Qualcomm and Omnipoint plan to offer chip sets implementing their technology. Qualcomm has developed its own chips for the cellular telephone industry, while Omnipoint announced an alliance with Rockwell International to put the DS-1900 in silicon. Omnipoint has also entered a 5-year, $100 million agreement with Northern Telecom for building and operating a PCS network in the New York area.

Of course, there is much more to the PCS technology debate than transmission

Figure 6.3 Omnipoint's family of spread spectrum–based PCS gear.

schemes. Network architectures and interfaces to the public telephone network are also hotly contested. Given the U.S.'s long tradition of voluntary standards and the successful track record of proprietary technologies, it is unlikely that vendors will agree on these matters anytime soon. While early standards may help a modest-sized market emerge in Europe, competition will guide U.S. players to achieve the price/performance ratios necessary to unleash the mass market.

6.3 PCS APPLICATIONS

If history is a reliable guide, the PCS will succeed in ways other than those anticipated by business planners. The microcomputer was supposed to blossom as a home computer; instead, the spreadsheet made it an essential desktop business tool. The first television broadcasts showed radio announcers standing in front of their microphones. Will the PCS really threaten cellular telephone's core business, or will applications we have not yet dreamt of emerge? Let's take a closer look.

6.3.1 Cheaper, Better Cellular

The idea of PCS as "cellular for the rest of us" may be alluring, but it is highly dubious. Microcellular networks operating at 1.8 GHz will most likely be more expensive to build and operate than 800/900-MHz cellular networks. In order to compete with cellular, PCS networks may have to achieve operating costs of 65% to 75% less than cellular's. In the United States, PCS operators will have to pay for their licenses and then pay to relocate incumbent microwave users.

Washington, D.C.–based American Personal Communications (APC) has suggested the first PCS networks must employ larger cells than previously assumed—perhaps approaching the size of cellular telephone cells. (APC won a Pioneer's Preference award for its frequency agile sharing technology (FAST), which integrates Qualcomm's CDMA technology and facilitates band sharing with incumbent microwave users.) When one considers estimates of $300 to $1,000 per subscriber to upgrade existing cellular networks to digital—more than the original cost of constructing the analog networks—the idea of competing head to head with cellular begins to look more interesting.

Telepoint's lack of success in the United Kingdom suggests personal communication services offering anything less than cellular telephone functionality are likely to have a tough time competing. Cellular networks in the United States offer near-ubiquitous coverage, handsets are available from multiple suppliers for under $100, service is available in a wide variety of packages, and many carriers now offer enhanced features such as voice mail, DMS, and even two-way data services.

All four of Britain's Telepoint services—Callpoint, Zonephone, Phonepoint, and Hutchison's Rabbit—have shut down. When Phonepoint terminated operation in late 1991, it had 3,000 base stations, but just 800 subscribers. Although Hutchison Personal Communications did not begin offering its Rabbit service until after the other three licensees ceased operations—giving Hutchison the entire market to itself—it attracted less than 10,000 subscribers and was forced to discontinue operation in late 1993. In France, where cellular penetration is relatively low, France Telecom's BiBop service has managed to attract over 25,000 users. But the lone success story is Hong Kong, a city-state in which three operators serve a total of over 150,000 users.

What went wrong with Telepoint in the United Kingdom? The three original operators felt the lack of an early CAI and initial reliance on low-quality British-made handsets were two critical mistakes. Indeed, Motorola's SilverLink became the handset of choice toward the end. (Motorola's PPS2000 PCS base station is shown in Figure 6.4.)

But the real problem was more fundamental. Telepoint had been positioned as a wireless callbox service. The old red callboxes were frequent targets of vandals and generally user-unfriendly. By the time Telepoint service was launched, sleek new credit card phones were popping up all over London, attracting consumers as well as business users. People did not subscribe to Telepoint because they had no need for a portable phone (typically costing $300) that could only make outbound calls from certain locations.

But let us assume there is a need for a portable phone that can place and receive calls, yet is less expensive to own and operate than a cell phone. Because interconnection to the public telephone network is a major chunk of the operating costs of

Figure 6.4 Motorola's PPS2000 PCS base station can be used for both fixed and pedestrian mobile services.

a mobile telephone service, companies with existing wireline networks (or even rights of way) could be well positioned to offer low-cost PCS—companies like LECs, alternative common carriers, cellular telephone operators, cable TV providers, interexchange carriers (IXC), and utilities.

An FCC Office of Plans & Policy Working Paper, "Putting It All Together: The Cost Structure of Personal Communications Service," concluded there are economies of scope between PCS and local telephone, cable TV, and cellular networks. (Economies of scope are realized when one network provides multiple services.) According to the study, PCS networks based on the existing infrastructure would realize a savings of $65 to $83 per subscriber per year over standalone PCS networks. Unfortunately, where some see the infrastructure to make PCS a viable business, others see an unfair advantage. Although cellular carriers are best qualified to offer low-cost cordless services, their participation in PCS has been restricted.

6.3.2 Personal Number Calling

These days, it is not uncommon to receive business cards indicating three or more phone numbers. As the demarcation between business and personal life dissolves and cellular telephone attracts more subscribers, it becomes increasingly difficult to know which number to use when trying to reach someone.

Some believe PCS is the solution to this dilemma. A cordless phone that works on the street, in the office, and at home suggests the need for one phone number that can be used to call the same person anytime, anywhere via an advanced intelligent network (AIN).

Personal number calling, however, need not wait for direct support by the PSTN. There are products and services on the market right now that offer "one-number calling." Telular has developed the PCS^{one}™ family of products, enabling cellular phone calls to be answered from standard (wired) phones in the home, office, on a boat, in a recreational vehicle, or from a vacation home. PCS^{one} is based on Telular's patented wireless-to-wireline interface technology. Of course, there is a tradeoff: the user must pay cellular airtime charges on every inbound call.

There are also special wireless phones that can be used in conjunction with telephone company AIN services, but do not entail airtime charges when operated indoors. Panasonic Communications & Systems Company has developed a PBX/Centrex adjunct called FreedomLink™ that provides one phone for both indoor and outdoor use. The handset is a specially programmed cell phone that defaults to the indoor system whenever it is within range. The indoor system consists of base station groups connected to a control unit that handles cell handoffs within the building and interfaces to the local PBX. A special scanning station continuously monitors the (outdoor) cellular network to determine which frequencies are free for use by the indoor system. Motorola has developed a similar product—the Personal Communicator 550—that serves as an N-AMPS cell phone outdoors and reverts to cordless phone operation when it senses the presence of a PPS-800 base station in the home or office.

Perhaps Arthur C. Clarke is right: in the future, we will accept nothing less than the ability to call individual people. But that does not mean you must be reachable anytime, anywhere, and by anyone. Accessline Technologies (Bellevue, Washington) provides a service dubbed Personal AccessLine™ that simplifies access for the caller, while giving the subscriber complete control over his or her accessibility. The AIN-compliant AccessLine System is a remote processor connected to a telephone company central office (or other switch) that interconnects landline, cellular, paging, voice mail, fax, e-mail, and other services to a single phone number. For urgent calls, AccessLine Smart Pager^{sm} instructs the subscriber to go to a telephone for immediate connection to the caller (a "meet me" service). Screened AccessLine^{sm}, on the other hand, permits the subscriber to listen as the system asks the caller to state his or her name; the subscriber can then decide whether to take the call or route it to voice mail. The subscriber may set up a computerized call-handling calendar, indicating where he or she can be reached throughout the week.

6.3.3 One Handset for Home, Office, and on the Street

Callers like the idea of one person, one phone number, because it simplifies access. Mobile users, however, may be more concerned with cost. While products from

Motorola, Panasonic, and Telular integrate dissimilar services, PCS promises coordinated frequencies and air interfaces. Public areas would be served in the licensed PCS band, while private areas would be served in the unlicensed PCS band. Because 20 MHz of unlicensed PCS bandwidth is located in the middle of the licensed PCS band, it should be easy to develop products that will work over both public and private systems.

But the idea of one phone for both business and personal use creates a new set of problems. It is difficult—if not impossible—to keep charges for personal and business calls separate. Most firms will pay a certain amount for personal calls while an employee is more than 50 miles from home on company business. It may be more difficult to place limits on inbound calls. As employees put in longer hours, personal calls made while at work will surely increase. And what happens when an employee loses a wireless handset? One major corporation has considered this policy: the company pays for the first handset an employee loses; the employee pays for each subsequent replacement.

6.3.4 Competition in the Local Loop

In the early 20th century—before the advent of modern switching systems and multiplexed circuits—streets in New York City began to grow dark from the sheer volume of overhead, crisscrossing phone lines. If there was ever a justification for local telephone company monopolies, this was it.

In hindsight, we know technology was the solution to the overabundance of aerial cables. Now, wireless technology is poised to remove all remaining obstacles to competition in the local loop. Wireless distribution systems avoid most of the environmental issues provoked by cable installations. NYNEX has been evaluating wireless local loop technology in anticipation that it will become less expensive than copper in certain applications.

Wireless local loop technology has other advantages. Wireless systems are easier to install and maintain than wired distribution systems. Wireless local loop technology is ideal for bandwidth-on-demand services because it is easy to add or subtract channels. Problems that plague buried copper loops—water, bridge taps, and the ravages of time—are eliminated. It also allows the LEC to get out of the power-generating business; power (including emergency backup) for wireless customer premises equipment (CPE) must be supplied by the customer. Certainly, wireless compares favorably to twisted-pair copper, but it cannot compete with fiber-optic cable for symmetrical, interactive, multimedia applications.

Ameritech has developed what it calls an ISDN-based ". . . open network interface for PCS, making possible a market for multiple providers and multiple support networks," and ". . . the development of new public switched telephone network (PSTN) capabilities and technologies to satisfy a broad range of mass market service needs." Ameritech emphasizes its open interface is wireless technology—

independent. In other words, this RBOC recognizes that local-access monopolies are doomed, and its switching network—rather than its copper loop plant—is its most valuable asset.

Interexchange carriers (IXCs) are also interested in PCS. They complain that local access fees paid to LECs account for nearly half of their costs. Not surprisingly, the U.S.'s three major IXCs—AT&T, MCI, and Sprint—are all pursuing wireless strategies. AT&T has vehemently denied its merger with McCaw Cellular Communications is an attempt to get back into the local phone business. In late 1992, MCI petitioned the FCC to establish three national PCS consortia, but later announced it was investing in a nationwide ESMR, Nextel. Sprint has acquired Centel, an independent phone company, along with its cellular operations.

But the plot thickens. Cable TV operators are also interested in the local telephone business. They believe they possess a strategic advantage over cellular carriers in delivering PCS in residential areas: their extensive coaxial cable plants. Companies like Cox Enterprises, Jones Intercable, Cablevision Systems, and the Warner Cable Group have been conducting PCS experiments.

For example, Jones Intercable plans to install PCS base stations in cable TV amplifier housings. These could be mounted on buildings and utility poles. Cablevision Systems Corporation has conducted experiments in both the ISM (using spread spectrum) and 12-GHz CARS (cable television relay service) bands. The firm claims wireless will make it easy to add two-way communications to their network. Cable TV Laboratories—the research arm of a cable TV consortium—believes cable TV networks employing radio-over-cable technology will prove an excellent platform for microcellular networks. (Radio-over-cable technology enables an entire radio band to be "pumped" over a coaxial or fiber-optic cable. Only a cable-to-radio converter and an antenna are required at the remote end.)

Cellular telephone carriers are interested in competition in the local loop, and they are already starting to provide it. In some markets, cellular is cheaper than local flat-rate wireline service for low-volume users (e.g., <140 min per month). No doubt, as cellular carriers migrate to higher capacity digital radio, they will become even more interested in luring customers away from LECs.

Thanks to the FCC Rules, there are no truly nationwide cellular carriers in the United States. PCS, however, gives carriers stymied by the cellular duopoly a second chance. While PCS entrepreneurs dream of becoming the third, fourth, or fifth "cellular carrier" in major cities, cellular carriers dream of using PCS to put together seamless nationwide networks. But to do that, they will need cheap dual-band handsets.

In the United Kingdom, PCNs have taken on a different role. They compete with cellular carriers for mobile service and with British Telecom for fixed services. Two firms are licensed to provide PCNs in the United Kingdom: Mercury Communications and Microtel. Mercury Communications' One-2-One service was launched in the Greater London area in September of 1993 and has acquired over 100,000 subscribers in its first year of operation. Although handsets were introduced at a

Table 6.1
Service Rates for Mercury Personal Communications' One-2-One PCN Service

	BusinessCall	*PersonalCall*
Initial connection	$31.00	$31.00
Monthly subscription	$31.00	$19.38
Peak-rate calls	$0.248/min	$0.3875/min
Off-peak, long-distance	$0.1240/min	$0.1550/min
Off-peak, local	$0.1240/min	Free
Calls between One-2-One users	$0.08/min	Long-distance rates
Inbound Calls		
Peak	$0.2371/min	$0.2371/min
Standard	$0.1736/min	$0.1736/min
Off-peak	$0.1116/min	$0.1116/min

Source: *Micro Cell News*, September 10, 1933, Vol. 4, No. 19, Probe Research, Cedar Knolls, NJ.

hefty $375 and above, One-2-One offers free off-peak local calling and peak-rate charges lower than those offered by the country's cellular networks. Mercury expects its PCN to become profitable within 4 to 5 years, but most analysts believe that to do so, the service must achieve 20% market penetration. One-2-One's introductory pricing is detailed in Table 6.1.

6.3.5 Wireless Multimedia

PCS providers are unlikely to beat cellular carriers on price, coverage, or smaller handsets. What they can beat cellular on, however, is providing enhanced services. Integrated voice and data, fax, graphics, still-frame video, and even full-motion video are important wireless opportunities.

Many observers believe wireless communications is best suited to narrowband applications. The well-known "Negroponte Flip" predicts telephone (i.e., narrowband services) and broadcast television (i.e., broadband services) will trade media, with the traditionally wired telephone becoming wireless and the previously wireless television broadcasting converting to cable. This has been an obvious near-term trend, but there has also been considerable activity in the development of broadband wireless services, suggesting that the Negroponte Flip may be wrong. For example, the FCC has proposed establishing a local multipoint distribution service (LMDS) in the 27.5- to 29.5-GHz band in response to a Petition For Rulemaking from the Suite 12 Group.

The Suite 12 Group's CellularVision™ is a fixed-point wireless system for two-way voice, video, and data—initially targeted at video-on-demand applications. Using cells 3 to 6 miles in radius, the service operates on 1-GHz-wide channels. Each cell is assigned either vertical or horizontal signal polarization for the 1-GHz video distribution channel, and the opposite polarization for the 1-GHz data uplink channel. By reversing polarization, the two channels may be reused in adjacent cells. The subscriber antenna is either a 3-inch window patch or a 15-inch dish.

In early 1994, Microsoft chairman William Gates and McCaw Cellular Communications chairman Craig O. McCaw announced a joint venture (called Teledesic) to pursue the construction of a worldwide low-earth-orbit satellite (LEOS) network to provide two-way interactive multimedia communications to fixed locations (including residential). The proposed network would consist of 840 satellites, operate in the 20-GHz band, and cost an estimated $9.55 billion to build.

The Teledesic announcement dovetails nicely with the Clinton administration's plans for a national information infrastructure (NII) with one key exception: Teledesic would enable a global information infrastructure (GII). Noting Vice President Albert Gore's concern that the United States not devolve into a society of information haves and have-nots, and taking great care not to raise the ire of LECs and cable TV operators, Teledesic is presented as strictly a rural and remote area solution.

What is exciting about Teledesic, however, is that it could upgrade the entire planet's telecommunications infrastructure in one fell swoop. While the estimated $9.55 billion construction cost may be too low, Teledesic would certainly cost less than the alternative: upgrading the worldwide terrestrial network. The idea that anyone would undertake such a project solely to serve rural and remote areas, however, stretches credulity.

But what about mobile users—how would they interface with multimedia services? There are at least a couple of interesting possibilities. For the vehicular

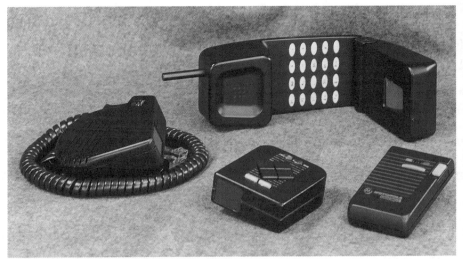

Figure 6.5 Reflection Technology's virtual display technology comes in a variety of form factors.

user, there is the color graphics display that serves as a sort of dash-mounted road sign. Instead of roadsides cluttered with billboards and message signs, advertisements and traffic alerts could be broadcast directly to automobiles. (More on this in Chapter 11.)

Reflection Technology (Waltham, Massachusetts) has developed what it calls a *virtual display* for cellular telephones. This is a miniature display held a few inches from the eye that provides the information contained on a full-sized high-resolution screen. The display uses a linear array of light-emitting diodes (LED) and a counterbalanced resonant vibrating scan mirror and magnifying lens. The mirror sweeps across the viewing field to create the perception of a full screen. The firm describes the combination virtual display/cellular telephone as "A phone that receives faxes. A phone that displays E-mail. A phone that accesses [personal computers] and database services" (Figure 6.5).

6.4 THE POLITICS OF PCS

PCS is at once an investment opportunity, a strategic global industry, and (apparently) the birthright of a large segment of American society. To speculators, PCS is the hottest wireless opportunity since the cellular telephone lotteries over a decade ago. No matter how regulators try to keep speculators out, they always find a way in. Rules prohibiting the near-term resale of PCS licenses will not do the trick—they will simply force speculators to outsource construction, operation, and maintenance to third parties.

To the government, PCS is the key to jobs at home and burgeoning markets abroad. Government officials believe they can hasten the emergence of a PCS industry through alliances with vendors. Whether it is mandatory standards, eligibility restrictions to ensure competition, or auctions to help balance the federal budget, PCS has become a colossal political football.

Many see the PCS licensing process as an opportunity to redress past grievances. A big chunk of PCS has been reserved for minority-owned businesses, woman-owned businesses, small businesses (redefined to include businesses with up to half a billion dollars in assets!), and rural telephone companies. These so-called *designated entities* will receive spectrum auction discounts; that is, they will not have to pay the full value of winning bids. The problem with this solution is that some of the entrepreneurs who pioneered PCS could find themselves squeezed out by Fortune 500 companies on one side and designated entities on the other.

The FCC Rules governing broadband PCS have created a maelstrom of controversy. Eligibility restrictions, the size of geographical service areas, bandwidth allocations, the spectrum auction process, and the fate of Pioneers' Preference awards have been hotly contested. More than 2,000 licenses will be awarded through auctions in 51 MTAs and 492 BTAs. Operators will be permitted to aggregate up to 40 MHz per market out of the available 10- and 30-MHz channels. It will be possible to obtain regional or national licenses if the amount bid exceeds the sum of the individual highest bids for the same licenses.

Cellular carriers are restricted to a single 10-MHz block in markets where they own at least 20% of a cellular property or currently serve more than 10% of the population. Licensees with 30-MHz blocks must offer service to at least one-third of the population in their service area within 5 years, and two-thirds within 10 years of obtaining their licenses. Licensees with 10-MHz blocks must offer service to 25% of the population in their area within 5 years.

Not surprisingly, the majority of potential broadband PCS bidders lack a cogent strategy. Like wild animals, many have responded by herding together with others of their kind. A cable TV consortium has been established with members such as Comcast Corporation, Continental Cablevision, Cox Cable, Tele-Communications, Inc., and Time-Warner Entertainment. American Wireless Communication Corporation was formed to represent the interests of small businesses, women- and minority-owned businesses, and rural telephone companies. PCS Action was created to speed the rollout of PCS. The National PCS Consortium represents independent cellular carriers, SMR operators, and cable TV companies.

Spectrum auctions were first suggested in the 1950s to ensure market forces played a role in the spectrum distribution process. Although not everyone wanted to abandon comparative hearings—proponents of so-called "modified" comparative hearings promised new, streamlined procedures—support within the U.S. Congress blossomed when lawmakers heard estimates of how much industry might be willing to pay for licenses. There was also the feeling that a private auction takes place every time a radio or TV license is sold. Perhaps stories regarding windfall profits gained by cellular lottery winners was the final straw. But opponents contend auctions favor those with deep pockets, make consolidation of licenses more difficult, and generally fail to take into account the public's interests.

Everyone agrees that consumers benefit from interoperability standards, but there is disagreement over how standards should be developed and promulgated. PCS standards are desired for two main purposes. First, users would like to be able to use the same handset on multiple networks—regardless of location. Second, there is thought to be demand for a handset that can be used at home, in the office, and in public areas.

In the United States, PCS standards are unlikely to be established soon. There are simply too many things that must be standardized: an air interface for voice and data services, PSTN interconnection, roaming and intersystem handoffs, and an extended numbering plan—to name just a few. More importantly, the industry is waiting for the market to indicate what is and is not important.

6.5 HAS NEXTEL BEATEN PCS TO THE PUNCH?

A funny thing happened on the way to the PCS gold rush. While PCS hopefuls scramble to become the "third" cellular carrier, Nextel has suddenly emerged with a rival plan—a plan based on existing licenses and free of costs associated with relocating incumbent users and participating in auctions. (See Chapter 4.)

Nextel's MIRS technology multiplexes six voice channels, each employing 4.2-kbps VSELP speech coding, onto a single 25-kHz-wide SMR channel running at a composite rate of 64 kbps. Motorola claims MIRS will deliver a fifteenfold increase in capacity over analog SMR.

But Nextel does not see itself as the third "cellular carrier." It sees itself as a provider of personal communications services. Like broadband PCS, Nextel is constructing a digital network from the ground up. The first Nextel handsets, designed and manufactured by Motorola, are integrated voice/paging terminals. Nextel's SMS, which accommodates messages of up to 140 alphanumeric characters in length, will serve paging and dispatching applications. Unlike conventional paging, SMS is a store-and-forward system with receipt acknowledgment. (In other words,

Figure 6.6 A Nextel handset: personal communicator in two-way business radio clothing?

the handset does not need to be on all the time to receive pages; the network will store and retransmit unacknowledged pages.)

Nextel 's initial focus is upgrading existing SMR dispatch users and winning over some of the estimated 13.5 million private radio users. Nextel's market research suggests that over half of private radio users also use cellular and paging services; Nextel can offer dispatching, mobile telephone, and paging in one package (Figure 6.6).

Nextel plans to deploy its digital mobile network in 45 of the top 50 U.S. markets. The firm estimates it will cost $2.5 billion to build its nationwide network. Nextel is quick to point out that PCS, in contrast, will not be a factor until at least 1997. Meanwhile, the firm introduced its service in the Los Angeles area in 1993, but has encountered serious interference and voice quality problems.

Perhaps most telling was MCI's eagerness to invest in Nextel. Originally, MCI devised an elaborate PCS strategy, calling on the FCC to designate three nationwide PCS consortia. But as it became clear that the AT&T-McCaw proposed merger would go through and that PCS licenses would be auctioned off piecemeal, MCI refocused its strategy on Nextel. Failing to reach an agreement with Nextel, MCI is once again in search of a personal communications strategy.

6.6 PERSONALIZED COMMUNICATIONS SERVICE

Personal communications is not a place in the radio spectrum or a specific service. It is a communications paradigm shift. PCSs will be provided by cellular telephone carriers as they migrate to digital microcellular networks and they will be provided by ESMRs like Nextel as they expand beyond traditional dispatching. PCSs will be provided by paging operators as they migrate from one-way to two-way messaging. And PCSs will be provided—but not exclusively—by licensees in the new 1.8-GHz PCS band.

PCS players are unlikely to beat cellular at its own game. But they do have the opportunity to introduce an array of new services: pocket fax machines, remote access to office resources, portable answering machines, and wireless newspapers. As long as PCS is dominated by those anticipating a government-driven gold rush or preferential treatment for special interest groups, it is almost certain to bring disappointment. The personal communications revolution will succeed by letting bold entrepreneurs test radical new ideas in the marketplace.

SELECT BIBLIOGRAPHY

[1]   FCC Office of Plans & Policy Working Paper No. 28, "Putting It All Together: The Cost Structure of Personal Communications Service," Washington, D.C., November 10, 1992.

[2]   Haynes, P., "End of the Line," *The Economist*, October 23, 1993.

[3]   *Microcell News*, Probe Research, Cedar Knolls, NJ, Vol. 5, No. 1, December 10, 1993, and Vol. 4, No. 19, September 10, 1993.

[4]   Raymond, K. R., NYNEX Science and Technology, "New Radio Technology for the Public Network," presentation at Cordless'90 conference, 5 December 1990, Miami, Florida.

[5]   Schnee, V., "Nextel's ESMR Arrives: A New Era of Cellular Integrated Services," *Wireless: For the Corporate User*, January/February 1994.

[6]   *Telecommunications Reports Wireless News*, Washington, D.C., issues throughout 1993/1994.

[7]   Tuttlebee, W., ed., *Cordless Telecommunications in Europe*, London: Springer-Verlag, 1990.

[8]   Varrall, G., and R. Belcher, *Data over Radio*, Mendocino, CA: Quantum Publishing, 1992.

CHAPTER 7
▼▼▼

THE WIRELESS ENTERPRISE

7.1 WIRELESS WIRE?

Wireless local-area networks (WLAN) were supposed to supplant much of the copper wire that snakes its way through modern offices. To the bitter disappointment of vendors, however, that has not happened. Despite disastrous results in the office market, WLANs represent an engineering triumph. As users demand untethered access to information, the high speeds and low battery drain of short-haul radio and infrared links will prove invaluable.

The first WLANs were designed to replace office cabling. Manufacturers hoped to ride the coattails of the multibillion dollar LAN industry. They convinced themselves that the cost of office cabling—when hidden expenses were considered—had become exorbitant. Although WLANs required significant initial investment, they would clearly save money over the long run.

Motorola went to great lengths to discover the true cost of cabling. Labor for installation and (later) moves, adds, and changes was identified as the biggest expense, particularly in major cities like New York. Although harder to measure, there are costs and frustrations associated with the time lost waiting for installations. Motorola even demonstrated that accommodations for cable—cable ducts and related structural support—can add 10% to the cost of a new building. Cable installation may also pose a safety hazard to office workers who must navigate past ladders and cables strewn over floors.

Nevertheless, offices have not embraced WLANs. If anything, much of the LAN market's growth can be traced to increased reliance on inexpensive twisted-pair telephone wiring. New buildings are often prewired, structured wiring systems simplify cable management, and city building codes have been revised to eliminate outdated requirements, like laws stipulating LAN wiring must run through conduits.

Indeed, the argument that WLANs save money is seriously flawed. Most moves, adds, and changes involve both a workstation and a telephone set. The cable technician does not care whether the wires carry voice or data; once a telephone has to be installed or moved, a workstation can usually be included at incremental cost. In large organizations—those most likely to benefit from WLANs—internal billing practices often obscure any savings by failing to differentiate between moves that require rewiring and those that do not.

More importantly, there is little incentive for customers in large organizations to purchase WLANs, which involve large up-front capital expenditures—often requiring multiple approvals. Cable installation charges, on the other hand, are treated as routine monthly expenses. Management is increasingly skeptical of technology that promises a long-term return on investment—especially technology they perceive as unproven, vulnerable to interference, and inherently insecure. (There are not only concerns about eavesdropping, but also airborne viruses and even intentional jamming.)

There are certainly niche applications for wireless office LANs. Cabling is not a practical solution for historic buildings, older buildings, buildings with asbestos insulation, and buildings that have simply run out of duct space. WLANs may make sense when workers are quartered in temporary offices. Wireless is also a good solution for bridging LAN segments between nonadjacent floors within a building, or between physically separated buildings.

Outside the office, there is a plethora of markets for WLANs, several of which are well developed. As firms begin to extend automation beyond the office, data collection (often in conjunction with bar code scanning) in warehouses and on factory and retail floors is burgeoning. Point-of-care systems in hospitals and clinics are also taking off. The growing base of portable computers may even lead to new applications within offices—not replacing cable, but providing portable access to desktop computers, file servers, printers, and fax machines. And mobile LANs are making a major push into business and university campus environments.

WLANs are, in fact, the next generation of mobile data technology. High-speed microcellular networks have crept out of the building and onto the campus. From there, they will gradually extend their reach throughout metropolitan areas. Perhaps through guerrilla warfare, WLAN vendors may one day challenge nationwide mobile data networks like those owned and operated by ARDIS and RAM Mobile Data.

7.2 PORTABLE DATA

Data collection in warehouses and on factory and retail floors is the largest WLAN market. Point-of-care automation in hospitals is growing. But the largest markets will probably encompass horizontal applications. One nascent application is "walkup" access to desktop PCs, printers, file servers, telephones, and fax machines. Another anticipated horizontal application is spontaneous and collaborative networking between personal digital assistants (PDA), personal communicators, mobile companions, and other portable devices. Let's examine some of the key WLAN markets.

7.2.1 Retail Inventory Tracking

Bar code scanning took off in the mid-1970s when low-cost laser scanners and microprocessor-based bar code printers were introduced. By the early 1980s, 90% of supermarket merchandise in the United States supported bar code–based automatic identification. Portable terminals are used to check stock and reorder and price goods on retail shelves. Companies like Kmart and Wal-Mart are using RF data collection systems to track shelf inventory, bypass warehouses, and reduce overall costs. See Figure 7.1. The U.S.'s shift towards a distribution-oriented economy was explained by management consultant Peter Drucker in an editorial in the September 24, 1992 issue of *The Wall Street Journal*:

> What underlies this shift is information. Wal-Mart is built around information from the sales floor. Whenever a customer buys anything, the information goes directly—in "real time"—to the manufacturer's

Figure 7.1 Wireless bar-code scanners from Symbol Technologies help track retail shelf inventory.

plant. It is automatically converted into a manufacturing schedule and into delivery instructions: when to ship, how to ship, and where to ship. Traditionally, 20% or 30% of the retail price went toward getting merchandise from the manufacturer's loading dock to the retailer's store— most of it for keeping inventory in three warehouses: the manufacturer's, the wholesaler's and the retailer's. These costs are largely eliminated in the Wal-Mart system, which enables the company to undersell local competitors despite its generally higher labor costs.

7.2.2 Hospital Point of Care

WLANs allow extending hospital information systems (HIS) to the bedside. Clinicom (Boulder, Colorado) sells a system that employs portable touch screen terminals as electronic patient charts (Figure 7.2). A number of hospitals are now bar-coding patients, health care providers, and medications to ensure that proper procedures are followed and to provide complete audit trails. Pen-based computers equipped with spread spectrum radios can also be used to instantly transmit prescriptions from the patient's bedside to the hospital pharmacy, enabling the pharmacist to respond with additional questions or choices (e.g., inform the doctor that a generic equivalent is available). In the emergency room, wireless terminals are

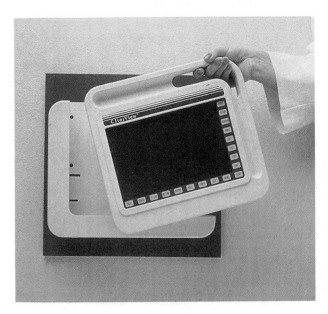

Figure 7.2 Portable touch screen terminals made by Clinicom serve as electronic patient charts in hospital point-of-care applications.

being introduced to assist physicians in documenting procedures and accessing lab results without having to leave the patient's side.

7.2.3 Financial Trading Floors

Alleged abuses of the 140-year-old "open outcry" system have led to trading pit automation efforts. Under the existing honor system, both the buyer and seller record the agreed-upon price and quantity; if they do not match, the traders do their best to reconstruct the deal at the end of the day. An automated system would ensure all trades are entered, time-stamped, and verified in real time. The goal is to replace or augment the trading card with a handheld terminal equipped with wireless data communications capability.

The Chicago Board of Options Exchange and the Chicago Mercantile Exchange have formed a joint committee called AUDIT to oversee the migration to an automated system. One interesting requirement is the ability to restrict communications to within a single trading pit, even though another pit may be located in the same open area. At first it appeared that the AUDIT committee was leaning, for security reasons, toward an infrared solution. But later it became apparent that application transparency—ensuring the traders can function much as they did before—was the top priority, since the committee selected a spread spectrum–based approach.

7.2.4 Point of Sale

For department stores that remodel on a frequent basis, WLANs permit easy reconfiguration of point-of-sale (POS) networks. Wireless also simplifies planning, eliminates unsightly cables, and enables POS terminals to be set up where cables often cannot reach—as may be required for the occasional sidewalk sale. In a related application, WLAN technology can be used at trade shows to link exhibitors to a centralized database containing information on attendees, who are identified by bar code, card swipe, or RF identification tags as they enter vendors' booths.

7.2.5 University Campuses

Twenty-five years ago, pocket calculators became required instruments for engineering students. Today, personal computers have become required tools for pupils in virtually all disciplines. The next phase may be the widespread adoption on campus of laptop, notebook, and palmtop computers. Wireless connectivity will be used to access online reference services for collaborative classroom computing and test administration and for rapid deployment of high-speed Internet links. The latter application is being pursued by companies like Metricom (see Chapter 3) and Tetherless Access (Fremont, California).

7.2.6 Enhanced Meetings

Tablet computers and "live boards" can provide a shared work space for meeting participants, helping them exchange ideas and jointly manage projects. The tablet is a pen-based computer with a screen of roughly the dimensions of a note pad. The live board is an electronic white board (i.e., a wall- sized display that one can write on using special electronic pens or (in remote control mode) wireless pen computers). Special software distinguishes input from different individuals, enables one person to (optionally) control what is displayed, and allows input to be erased by sequence, source, or other criteria.

At the end of an electronically enhanced meeting, each attendee retains an identical copy of the meeting notes. (Each user is also able to annotate his or her copy with private comments.) Sketches and ideas can be recalled later by individuals or groups, or shared with others who were not present. Apple Computer has even suggested that people might participate in meetings electronically—via remote access to live boards—perhaps enabling them to attend two meetings at the same time!

7.2.7 Personal Data Interchange

As PDAs and other pocket information appliances proliferate, people will shift much of their data storage from paper to electronic media. PDAs will replace the conventional printed business card: users will "beam" (or "squirt") virtual business cards from PDA to PDA via infrared links. Users will also exchange data between handheld and desktop computers. In another envisioned application, PDAs will one day communicate via infrared with pay phones to provide dialing instructions, download text summaries of voice mail messages, and exchange multimedia postcards.

7.3 WIRELESS LAN TECHNOLOGIES

There is still some disagreement over the definition of WLANs. Purists argue that WLANs must either be compatible with industry-standard office LANs—Ethernet, Token Ring, or Arcnet—or support peer-to-peer communications (a feature most wireless data collection networks lack). A more generous definition is any wireless data network designed for use over short distances (<1 mile).

The first WLANs were sub-LANs; that is, they ran much slower than industry-standard LANs. They were also called *zero-slot* LANs because they used existing RS-232 serial ports instead of PC add-on slots. Wireless zero-slot LANs are primarily used for peripheral sharing and occasional PC file transfers. They generally work on a point-to-point contention basis. In other words, although there may be several PCs and printers on the network, it is only possible to establish one point-to-point connection at any given time. Later, WLANs were introduced based on special network interface cards (NIC) and software for compatibility with industry-standard LANs.

In early 1991, Motorola (Schaumburg, Illinois) introduced the first WLAN claiming to be both compatible with and transparent to standard Ethernet LANs. Motorola's Altair was billed as "wireless wire" because it plugs into existing NICs and replaces only the Ethernet cable (Figure 7.3). Altair has achieved limited success—in part due to its hefty price tag (initially, just under $7,500 to connect six workstations to an Ethernet backbone). Subsequently, other firms have also introduced WLANs designed to replace Ethernet or Token Ring LAN cabling.

Radio frequency/data collection (RF/DC) systems—although not initially recognized as WLANs—have achieved the most success. The first radio frequency/identification (RF/ID) systems were configured as star networks, ran at low speeds, and employed host-driven polling protocols. By and large, these products have migrated to unlicensed spread spectrum radio technology developed for wireless office LANs, and they have replaced their polling protocols with more efficient random-access techniques.

(In a polled network, terminals are only allowed to transmit when granted permission by a host computer. Typically, the host broadcasts terminal addresses in sequence, briefly pausing for responses after each. Polling is efficient if most of the

Figure 7.3 Motorola's Altair is a desktop replacement for Ethernet cabling.

terminals are active most of the time. If not, time is wasted polling terminals with no pending transactions. In a random-access network, in contrast, any terminal may transmit as long as another transmission is not in progress. Random access is more efficient when most of the traffic is bursty.)

Today's WLANs use five main transmission technologies: (1) narrowband FM radio, (2) low-power radio, (3) spread spectrum radio, (4) 18-GHz microwave radio, and (5) infrared lightwave. Each is defined as much by its regulatory status as its technical properties. All five are examined below.

7.3.1 Narrowband FM

Narrowband FM radio was once the most popular wireless connectivity solution for handheld data collection terminals. These terminals are often used in conjunction with automatic identification devices such as bar-code scanners, magnetic wands, or optical character recognition readers. Narrowband FM is simple, inexpensive, and generally performs at speeds of up to 9,600 bps.

In the United States, narrowband FM terminals generally operate in the 450- to 470-MHz band on licensed 12.5- or 25-kHz-wide channels. (It typically takes about 60 days to obtain the necessary license.) Transmit output power is limited to a maximum of 2W under FCC Rules Part 90, Subpart J, "Non-voice and Other Specialized Operations." Frequency allocations may be different or nonexistent in other countries, or, as in the case of Japan, transmit power may be severely restricted.

Generally called RF/DC systems, these products usually employ proprietary communications protocols between the portable terminals and fixed base stations. Data integrity is ensured through the use of an ARQ technique. A special computation (a CRC) is performed on each block of user data; the result (which tends to be unique for each block) is appended to the end of the block prior to transmission. Upon receiving the block, the receiver performs the same calculation and checks to see if its answer matches the one received. If it does not, it means that there is an error in the received data; the receiver instructs the transmitter to resend the block.

The advantage of narrowband FM is that the radios are small, inexpensive, and offer good range (both indoors and outdoors). The disadvantages of narrowband FM are that it is vulnerable to multipath, cochannel, and adjacent-channel interference, and the user must obtain an operator's license. (Each system is licensed to operate on a specific frequency at a specific location to protect users from harmful interference.)

Customers complain that licensing is a hassle and often fails to provide meaningful protection. Because they are licensed to operate in specific locations, most narrowband FM radios cannot be readily moved. Car rental companies complain that, despite licensing, interference remains a major problem in locations like airports, where there are many diverse users. Most of the major RF/DC vendors have now developed unlicensed spread spectrum versions of their products.

7.3.2 Low-Power Radio

Devices that transmit at very low power levels (typically on the order of 1 mW) may be manufactured for unlicensed use under FCC Rules Part 15.249. These devices are usually suitable for communications over very short distances—typically less than 100 feet.

Operation is permitted in the 902- to 928-MHz, 2,400- to 2,483-MHz, 5,725- to 5,850-MHz, and 24- to 24.25-GHz ISM bands. The ISM bands were created to accommodate devices that use radio frequency energy for purposes other than communications (e.g., medical diathermy machines, ultrasonic machines (such as humidifiers), and microwave ovens). As explained in Chapter 2, the use of Part 15 devices is permitted on a secondary basis only, and manufacturers must have their designs certified.

The primary advantages of low-power radio are that it is inexpensive, signals are easily confined within a building, and the devices may be used without obtaining an operator's license. The disadvantages are limited range and vulnerability to interference and multipath fading. Performance can be improved through the use of dynamic channel allocation (DCA) and/or antenna diversity. In a DCA system, the best frequency at each moment is automatically selected. Antenna diversity—the ability to switch antennas at one or both ends of the link—may also be used to combat multipath fading. Both of these techniques, however, add complexity and cost.

7.3.3 Spread Spectrum Transmission

Spread spectrum is by far the most popular WLAN technology. One reason is that FCC Rules Part 15.247 permit unlicensed spread spectrum operation at up to 1W of output power. Not surprisingly, many products barely meet the spreading requirements.

Most spread spectrum WLANs operate in the 902-MHz ISM band in North America, but many new products target the less crowded 2.4-GHz band. (In Europe, WLANs must operate in the 2.4-GHz band, since GSM cellular is located at the low end of the 900-MHz band.) There is little activity in the 5.725-GHz band; one exception is Windata's FreePort WLAN, which features dual-band (2.4/5.725 GHz) operation.

FCC Rules permit frequency-hopping, direct sequence, and hybrid spread spectrum systems. Frequency hopping is allowed in the 902-MHz band with a minimum of 50 hopping channels, and in the 2.4- and 5.725-MHz bands with a minimum of 75 channels. A frequency hopper must hop in a pseudorandom fashion so that, on average, each frequency receives equal use. The average occupancy time on a given channel must not exceed 0.4 sec over a 20-sec period in the 902-MHz band and 0.4 sec over a 30-sec period in the 2.4- and 5.725-GHz bands. In 1990, the maximum hopping channel bandwidth was increased to 500 kHz, opening the door to high-speed slow hoppers.

Direct sequence systems must possess at least 10 dB of processing gain. Part 15 defines processing gain as the ratio in decibels of the SNR with no spreading to the SNR with spreading as measured at the receiver. Hybrid systems are defined in terms of both frequency-hopping and direct sequence components; that is, if either component is turned off, the remaining component must meet the applicable rules.

Much of the interest in WLANs (and WPBXs) in the United States has focused on the ISM bands. However, five different priority levels are assigned: (1) true ISM devices, like industrial plywood dryers (highest priority), (2) government radiolocation (naval radar), (3) automatic vehicle monitoring (AVM), (4) amateur radio, and (5) Part 15 devices like WLANs (lowest priority). Users must not cause harmful interference to higher priority devices. Therefore, a WLAN may be forced to shut down if it causes harmful interference to a higher priority user.

It is more likely, however, that a WLAN in the 902-MHz band will receive harmful interference from a higher priority user. For example, AVM base stations are permitted to transmit at up to 1,000W. The 2.4-GHz band is becoming increasingly popular because it is less crowded (with the notable exception of microwave ovens), is available in Europe, and offers greater bandwidth. While the 5.8-GHz band is free of high-power transmitters and offers 125 MHz of bandwidth, it exhibits low-power efficiency and high equipment costs. In Europe, the 5.8-GHz band has been allocated to intelligent vehicle highway systems (IVHS).

7.3.4 18-GHz Microwave

Motorola's development of the Altair® WLAN operating at 18 GHz was an engineering tour de force. So far, however, the Altair Group's dream of wireless inbuilding networks (WIN) has proven elusive. Perhaps this product will eventually find success in point-to-point, curb-to-home applications.

Believing users would prefer the protection afforded by licensed operation, Motorola selected the 18-GH digital termination service (DTS) and convinced the FCC to modify Part 94.65 of the private operational fixed microwave service (POFMS) in 1990 to permit low-power inbuilding use. Motorola then proceeded to grab up 18-GHz licenses, consisting of two 10-MHz-wide channels each and providing exclusive use over the area within a 17.5-mile radius. In other words, Motorola would be the licensee—with Altair customers operating under its auspices.

Altair's radio link runs at 15 Mbps—fast enough to promise transparent operation for 10-Mbps Ethernet LANs. In order to maximize signal strength and minimize multipath interference, an intelligent six-sector antenna dynamically selects the best path between transmitter and receiver several times per second. Low-power operation at 18 GHz affords excellent security. Proprietary VLSI technology dramatically reduces the size and weight of the 18-GHz electronics. But high cost and limited range have conspired to restrict Altair's market.

7.3.5 Infrared

Amazingly, the first infrared WLANs targeted the same application as their radio-based cousins: cable replacement. Indoors, infrared LANs have demonstrated performance rivaling that of cable. Nodes are usually installed on masts to avoid (or minimize) interruptions from people walking through the beam path. Because it does not penetrate walls, floors, or ceilings, and is greatly attenuated by most window glass, infrared provides excellent privacy and access security. Infrared is also immune to RF interference. (However, infrared transmission is vulnerable to interference from other light sources, such as the sun.)

Outdoors, rooftop-to-rooftop infrared links may reach distances greater than 1 km, but are subject to occasional weather-related outages. Infrared is often used as a telephone company bypass solution in Europe, where many countries prohibit private microwave links. Indoors, range varies from a few inches (e.g., palmtop computer-to-portable printer) to in excess of 100 feet (point-to-point links between fixed nodes).

But mobile applications—particularly "walkup" data access—represent the largest opportunity. Infrared LAN vendors hope to leverage low-cost components developed for the TV remote control industry. While consumers do not mind pointing their TV remote controls, some vendors believe business users will demand omnidirectional operation. This has proven a stumbling block because diffuse (omnidirectional) operation requires significantly greater power than the point-and-shoot approach. Photonics Corporation (San Jose, California) has pioneered omnidirectional infrared transceivers with products such as its Collaborative™ PC (Figure 7.4).

Most PDAs, personal communicators, and mobile companions come equipped with infrared ports. Infrared is expected to provide a standardized means of exchanging data between PDAs or between PDAs and desktop PCs. Short-range point-and-shoot infrared is extremely inexpensive (approximately $1.50 to $4.00 in parts) and consumes very little battery power.

The Infrared Data Association (IrDA) was formed in 1993 to establish a worldwide cordless data connectivity standard and today boasts members representing over 50 computer and communication hardware, software, and component manufacturers. IrDA released its interoperable infrared serial data link standard (supporting speeds up to 115.2 kbps) in early 1994. IrDA emphasizes that none of the world's governments restricts the use of LED-based infrared communications. Association members include Apple Computer, AT&T, Casio, General Magic, Hewlett-Packard, IBM, Motorola, Northern Telecom, and Traveling Software.

In a press release dated 7 November 1993, IrDA saw compelling walk-up data transfer applications involving:

> . . . docking and input units, printers, telephones, desktop/laptop PCs, network nodes, ATMs, and mobile peers (PDA meets PDA) . . . For

Figure 7.4 Photonics' Collaborative family of infrared transceivers are used to create "instantaneous" indoor LANs.

example, users could quickly print a document, exchange business cards between handheld PCs, fax or e-mail directly from a notebook PC through a public phone, or store banking records from ATM machines by making a simple, point-and-shoot connection. Industrial and service applications find mobile IR devices enhancing control, documentation and docking procedures.

Infrared does have its disadvantages. Most low-cost infrared data links will not work in direct sunlight; this could prove a minor irritation in "PDA meets PDA" applications. In addition, mobility is somewhat limited because infrared does not penetrate walls. The buildingwide infrared solution is a hybrid wired/wireless LAN. Infrared bubbles (hemispheres containing arrays of infrared transducers) may be interconnected via an industry-standard wired LAN. This type of hybrid network, providing buildingwide mobility, has been developed by Spectrix Corporation (Evanston, Illinois) (Figures 7.5 and 7.6).

Figure 7.5 The Spectrix infrared access "bubbles" enable organizations to construct buildingwide portable data networks.

Figure 7.6 A handheld computer equipped with the Spectrix battery-powered infrared module.

7.4 PCMCIA CARDS

For years, the portable computer industry was plagued by the lack of a standard add-on card. Each brand, and often model, of laptop computer required its own proprietary cards. Not surprisingly, prices were high and selection was limited. The emergence of a standard plug-in card would seem essential to the future success of radio modems. PCMCIA WLAN cards have been introduced or announced by companies such as Proxim, AT&T, Motorola, and Xircom.

The Personal Computer Memory Card International Association was founded in 1989 to define standards for credit card–sized plug-in cards. Although it was originally defined for add-on memory, the PCMCIA standard was adapted to other requirements (such as communications) with the release of Version 2.0 of the PCMCIA PC card specification in September 1991.

There are three types of PCMCIA PC cards: Type I, Type II, and Type III. Each uses a thin 68-pin edge connector. The PCMCIA standard has been enhanced to permit "hot swapping" (i.e., inserting or removing a card while the host is on) and execute in place (XIP) (the ability to run computer programs on the card independently). Input/output PCMCIA PC cards will include dial-up modems, radio modems (both local and wide area), fax modems, LAN adapters, and docking station interfaces (Table 7.1).

Integrating wireless transceivers with portable computers via standard plug-in cards poses a number of challenges, including transceiver power, meeting FCC Part 15 requirements, and antenna (or infrared transducer) placement. To avoid missing a message, a pager card should be on all of the time; the solution may be a separate onboard battery. Wireless transceivers may also draw more power than their host. In some cases, the best solution is a two-piece transceiver: a plug-in card plus a self-powered outboard final amplifier module.

Part 15 requirements may be particularly problematic. By themselves, signals generated by the plug-in transceiver and portable host may be within FCC specifications. When the two devices are combined, their signals may mix, resulting in new, out-of-spec emissions.

Table 7.1
PCMCIA PC Card Form Factors

	Length	*Width*	*Height*
Type I	85.6 mm (3.37 inches)	54 mm (2.126 inches)	3.3 mm (0.13 inches)
Type II	85.6 mm (3.37 inches)	54 mm (2.126 inches)	5 mm (0.197 inches)
Type III	85.6 mm (3.37 inches)	54 mm (2.126 inches)	10 mm (0.394 inches)

Figure 7.7 Proxim's RangeLAN2/PCMCIA was the first credit card–sized WLAN adapter on the market.

Positioning the antenna or infrared transducer for optimal performance, safety, and nonobtrusiveness may also be tricky. The PCMCIA standard does not address where the slot should be located and how it should be oriented. This may be a problem if the antenna or infrared transducer is fastened to the edge of the card. Proxim (Mountain View, California) was the first vendor to ship a PCMCIA WLAN card; their solution is a separate antenna unit connected to the card via a short cable (Figure 7.7).

Ironically, PCMCIA communication cards still do not offer universal compatibility. Although the form factor, physical connector, and electrical interface have been standardized, the software interface has not. Special software drivers are required for each operating system and hardware platform. The ideal solution may be a universal software module that automatically uploads the correct driver whenever the card is inserted in a host.

7.5 UNLICENSED PCS

Despite fears of future overcrowding and the low priority assigned such devices, most WLANs and PBXs sold in the United States operate in the ISM bands. Understandably, vendors would prefer a band created exclusively for wireless office systems. Apple Computer deserves credit as the first major player to petition the FCC to set aside part of the new PCS for unlicensed wireless data devices.

In early 1991, Apple petitioned the FCC to establish a *data-PCS* based on a radio etiquette—a random-access technique that prescribes a minimum listening period before transmitting, a maximum transmission duration, and a minimum waiting period between transmissions. This etiquette would enable incompatible products from diverse vendors to coexist in the same band. More specifically, Apple proposed allocating 40 MHz in the 1,850- to 1,990-MHz band, citing the rapid growth of laptop and notebook computer markets, the evolution toward "spontaneous, collaborative computing" in the workplace and classroom, and the need to foster innovation to enhance U.S. global competitiveness.

This was not a battle Apple could win by itself. The Wireless Information Networks Forum (WINForum) was founded in 1992 (largely through the efforts of its first executive director, Benn Kobb) to promote the allocation of spectrum for "user-provided wireless voice and data personal communications services" (*user-PCS*). The organization, whose members include Apple, AT&T, DEC, Fujitsu, Hewlett-Packard, IBM, Sun Microsystems, and Tandem Computers, submitted a detailed etiquette proposal in May 1993. Although the organization had hoped to integrate wireless voice and data in the same spectrum allocation, members were unable to resolve their differences and settled on a service consisting of two subbands.

Meanwhile, the Unlicensed PCS Ad Hoc Committee for 2 GHz Microwave Transition and Management (UTAM) was formed to manage the relocation of existing microwave users—many of whom will require compensation and assistance. According to UTAM, it will not be possible to support true nomadic devices until all incumbent users have been cleared from the band.

7.6 WLAN STANDARDS

The weak market for wireless office LANs is often blamed on the lack of an industry standard. There is little evidence of this. For example, wireless data collection systems succeeded in the absence of a standard. Many highly regarded standards have failed to create markets, and many successful markets possess standards that evolved out of proprietary solutions. *Standards do not create markets; markets create standards.*

Standards ensure a wide selection of products that do essentially the same thing—a valuable role once a market begins to mature. There are two major WLAN standards efforts under way.

In the United States, the IEEE is developing WLAN standards. The IEEE, which already has over 600 standards to its credit, usually submits its draft standards to the ISO for worldwide ratification. Project 802 develops LAN standards and, in 1990, formed the 802.11 Wireless Access Method and Physical Medium Committee.

The committee defines *local area* as a building, campus, or outdoor storage/parking area. An important goal of Project 802 is making alternative physical media and access methods look the same to applications software; for example, an electronic mail program should not need to know whether a LAN uses twisted-pair wiring or radio.

The 802.11 committee is focusing on the 2.4-GHz band because it is relatively uncrowded and offers high bandwidth. The year of 1993 was important, since vendors began forming alliances around concrete proposals. A media access control (MAC) layer proposal was adopted with an ad hoc peer-to-peer system winning over a centrally managed approach. Both frequency-hopping and direct sequence 1-Mbps spread spectrum physical layer proposals were also approved. But even by the most optimistic estimates, a formal 802.11 standard is not expected before mid-1996.

In Europe, the CEPT spent 3 years defining cordless data features for what was primarily a cordless telephone system—the DECT standard. The 1-Mbps system, targeted at the 1.88- to 1.90-GHz band, was almost immediately judged obsolete, and work continues within the ETSI on a 10- to 20-Mbps High-Performance European Radio LAN (HIPERLAN).

7.7 WIRELESS TELEPHONE SYSTEMS

Although the term *wireless PBX* is widely used, only wireless extensions, adjuncts, and Centrex service are actually available. (Centrex is a telephone company service that supports many of the features of a PBX). Few organizations want or can afford an entirely wireless PBX. At best, wireless phone systems cost twice as much as comparable wired phone systems.

In large organizations, wireless telephone systems enable telecommunications managers to quickly set up phone service for new users, reach users in challenging locations (e.g., the factory floor), and provide a reliable means of accessing individuals who spend considerable time away from their desks. (Interestingly, it is the executive's assistant, rather than the executive, who needs to remain accessible.) Large organizations also see wireless phone systems as a vehicle for reducing information lag time. Engineers outfitted with wireless phones would not have to shuttle back and forth between their offices and the factory floor as frequently as they do now.

In small organizations, WKTSs may offer the best solution. (Key telephones usually feature a row of push buttons for selecting among a small number of lines.) With a wireless phone, retail salespeople do not need to put the caller on hold while checking to see if an item is in stock. They can also answer follow-up questions without having to run back and forth to a fixed phone. A WKTS not only enhances service to the caller, it saves time for the salesperson.

Wireless access can also be provided as a service on private premises. Typically, customers use dual-mode handsets to communicate with dedicated indoor base stations. When the user leaves the building, the handset automatically switches to the public cellular telephone network. In some cases, special low-power cellular base stations are installed indoors, enabling the customer to use a slightly modified cell phone as the dual-mode handset.

But the biggest opportunity for wireless telephone systems may be providing cordless pay phone service in airports, shopping malls, and sports complexes (perhaps using rental phones), and low-cost cellular service in residential areas, on campuses, and in downtown business districts (perhaps using personal cordless phones that also work at home). These services will be particularly attractive in fully competitive, local-access markets. (Sure, it would be tough for a new player to compete with the local telephone company and cellular carriers throughout their service areas, but it would be much easier to compete in selected high-traffic locations.)

Ericsson Radio Systems has suggested that it will develop a dual-mode wireless telephone. A vehicle-mounted base radio will work over a cellular telephone, while a detachable handset will work over private micro- or picocellular networks. In the office, the handset is used by itself. In the automobile, the handset plugs into the vehicle-mounted radio or can communicate with the base radio (which serves as a relay station) via a short-range radio link.

Just as mobility was found to be the principal benefit of WLANs, mobility is the main benefit of wireless phone systems. For example, medical specialists can be reached instantly anywhere within a hospital. On a campus, security and maintenance personnel can be quickly located and contacted. But unlike WLANS, wireless telephone systems have the ability to reduce phone bills. In large organizations in the United States, it is estimated that two-thirds or more of inbound calls fail to reach the called party directly. Using a system that rings the subscriber's desk phone and pocket phone simultaneously, few calls would be missed. Wireless telephone systems could also be used to eliminate pagers and public address systems or, in conjunction with these systems, enable key personnel to respond more quickly. Telephone tag is a poor use of a high-priced manager's time.

Many firms have instituted "hoteling" for consultants, auditors, and field sales representatives. These workers are paid to spend their time in the field with customers. If they are doing their jobs, they should not need permanent office space. When they must spend time in the home office, a wireless phone enables them to access their assigned extension from any desk that happens to be available.

WPBXs are even less likely than WLANs to succeed as a cable replacement. Cabling is still needed to connect the base stations, which serve two to eight simultaneous conversations each. More importantly, research suggests most organizations expect to limit cordless access to 5% to 15% of users. They are not interested in giving every employee a phone that can be lost or stolen. Totally wireless PBXs are also unlikely to replace today's feature-rich and highly reliable wired PBXs.

A different sort of wireless telephone system may be used in disaster recovery applications. Instead of giving users wireless handsets, wireless trunks may be used to bypass local landline service in the event of a telephone company central office or cable plant disaster. For example, Telular supplies cellular radios that connect to the trunk side of a PBX (the outside phone line side). In the event of a local service disruption, the PBX reroutes authorized users to the radios, which automatically access (prearranged) cellular service (assuming local cellular service has not also been disrupted).

Vendors, rather than market researchers, tend to uncover the most unexpected applications. For example, SpectraLink (Boulder, Colorado) reports that cigarette smokers represent an important niche market for wireless telephone systems. A growing list of firms restrict smoking; some are equipping smokers with wireless handsets so that they will not miss calls during trips to designated smoking areas.

7.8 WIRELESS TELEPHONE SYSTEM TECHNOLOGIES

The first generation of wireless phone systems were designed to interface with standard analog ports. Therefore, these systems could be sold as add-ons to virtually any brand of PBX. A potential drawback, however, is that analog ports do not always support all of the digital desktop phone features. A special radio controller is required to support mobility (i.e., manage cell handoffs) and enable features such as three-way calling (Figure 7.8).

Wireless phone systems have been promised that will work with digital ports. Since the PBX-to-phone digital interface has not been standardized, these systems will work exclusively with the proprietary systems for which they are designed. The

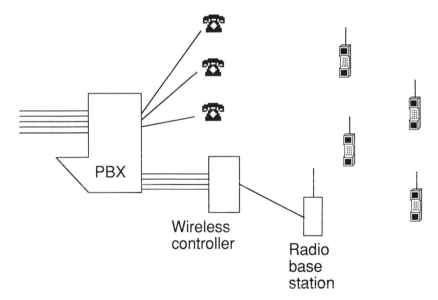

Figure 7.8 Wireless adjuncts permit organizations to equip selected users with wireless phones for use within the business facility.

advantages of digital ports are (1) cordless users will enjoy the same features that are found on digital desk phones, (2) mobility functions may be handled by the PBX mainframe, eliminating the need for separate wireless controllers, and (3) end-to-end digital transmission may be employed.

There are several competing radio interfaces for wireless phone systems, including CT1, CT2, CT2Plus, CT3, DECT, direct sequence spread spectrum, and CDMA. In some countries, frequencies have been allocated specifically for wireless office telephone systems. In the United States, wireless phone systems currently operate in the ISM bands. While ISM products have enjoyed modest success, we are unlikely to see end user organizations invest large sums until exclusive spectrum becomes available.

AT&T has introduced cordless extensions for the Merlin, System 25, Partner, and Partner Plus office telephone systems based on 46/49-MHz analog cordless technology (FCC Rules Part 15.233). The company is also developing a frequency-hopping spread spectrum solution, the TransTalk 9000 adjunct for the Partner, Merlin, and Definity PBXs. Preliminary specifications indicate that TransTalk 9000 will operate in the 902-MHz band. Each communications link will use 50 of 173 available (150-kHz-wide) hopping channels, for a total of up to 40 simultaneous conversations per system. AT&T believes that with a maximum portable-to-base range of 500 feet, TransTalk 9000 will not need to support cell handoffs.

Northern Telecom has developed the CT2/CAI-compliant Companion WPBX operating in the 864- to 868-MHz band. In the United States, these frequencies may only be used under an experimental license. Northern Telecom hopes to upgrade this product to the unlicensed PCS band.

The installation of a WPBX is as much an art as a science. Providing continuous radio coverage within a building poses unique challenges. Unlike cellular telephone networks, which are two-dimensional, wireless telephone networks may be three-dimensional. But how should users in stairwells or moving elevators be served? A user who enters an elevator on the first floor and pushes the button for the 49th floor may travel through dozens of cells in just a few seconds. The solution may be a dedicated cell in each elevator. But confining coverage to within an elevator (not to mention connecting a mobile base station to the PBX mainframe!) are additional problems.

7.9 WIRELESS TELEPHONE SYSTEM PIONEERS

Bell Atlantic Mobile Systems (BAMS) and Motorola have teamed up to offer a phone that works in the home, office, and automobile. Motorola's PPS 800 handset is a dual-mode unit for 49 MHz (residential cordless) and 800 MHz (cellular). It automatically recognizes and defaults to the 49-MHz base station whenever it comes within range. For its part, BAMS is supporting personal number calling; that is, mobile users may be reached via the same phone number whether they are in range of the cordless base station or operating over the cellular telephone network.

Ericsson Radio Systems is arguably the leading WPBX pioneer, having developed CT3. For the U.S. market, Ericsson modified its TDMA/TDD system (the model DCT 900) to meet FCC Part 15 requirements for operation in the 900-MHz ISM band. This product is marketed under the brand name Freeset by Ericsson and as the RolmPhone 900 by PBX maker Rolm Corporation. According to Ericsson, only a tiny reduction in the DCT 900's transmit output was required to conform with Part 15. Users can make and receive calls within 50 to 130 feet of a base station; handoffs between base stations are supported. Four radio carriers, each 1 MHz wide, are divided into 16 time slots—2 per full-duplex speech channel. The intelligent handset uses continuous dynamic channel selection (CDCS) to ensure optimal performance (Figure 7.9).

Southwestern Bell Telephone Corporation is offering an indoor wireless telephone service, called the FreedomLink Personal Communications System, in conjunction with Panasonic. FreedomLink is based on a low-power cellular base station using special control channels, which may either be connected to a PBX or a Centrex switch. The handset defaults to the low-power base station whenever it detects that it is within range. Outdoors, it operates like an ordinary cell phone. The base station works in conjunction with a frequency scanning station which maintains a list of

Figure 7.9 Ericsson's Freeset is based on CT3 technology and has been modified to meet FCC Part 15 regulations for sale in the United States.

unused local cellular channels. (The low-power indoor base station borrows unused cellular channels from the local cellular network.)

The SpectraLink Pocket Communications System (PCS) 2000 consists of a master control unit (MCU), remote cell units (RCU), and pocket phones (PT). SpectraLink claims the system supports up to 720 pocket telephones and over 400 simultaneous calls, and can cover an area in excess of 3 million square feet. The PCS 2000 uses a combination of TDMA/TDD (to separate calls) and CDMA (to isolate adjacent cells) in the 902-MHz ISM band (unlicensed operation). The pocket telephones operate in three modes: all calls ring through, no calls ring through, and selected calls (programmable code) ring through. The PCS 2000 is an add-on to any PBX or Centrex system and is designed for smaller installations, with a maximum capacity of 40 phones and 18 RCUs.

7.10 THE UNTETHERED ENTERPRISE

So far, neither WLANs nor WPBXs have achieved much success. But they will. These products represent an early foray into microcellular and picocellular networking. As they slowly proliferate, WLANs and WPBXs will offer untethered access to a wide variety of personal communications services.

Ultimately, these technologies will emerge as metropolitan and even wide-area solutions. Products that work both indoors and outdoors, over both private and public systems, will become commonplace. WLAN technology, particularly infrared, will enable spontaneous networking at meetings, conferences, and trade shows. We may never beam ourselves up to the starship Enterprise, but we will surely beam data back and forth between pocket information appliances and wireless business enterprises.

SELECT BIBLIOGRAPHY

[1]   Brodsky, I., "Is a Wireless PBX or Key System in Your Future?" *Business Communications Review*, April 1991.
[2]   Brodsky, I., "Introducing Wireless LANs," *Business Communications Review*, May 1990.
[3]   Brodsky, I., "Motorola and NCR Boost Prospects for Wireless LANs," *Business Communications Review*, January 1991.
[4]   Calhoun, G., *Wireless Access and the Local Telephone Network*, Norwood, MA: Artech House, 1992.
[5]   Chen, K.-C., "Medium Access Control of Wireless LANs for Mobile Computing," *IEEE Network*, September/October 1994.
[6]   Kaufman, J., "Wireless Telephone Systems—Not Wireless and Not a PBX!" *Business Communications Review*, June 1994.
[7]   Mathias, C. J., "New LAN Gear Snaps Unseen Desktop Chains," *Data Communications*, March 21, 1994.

CHAPTER 8
▼▼▼

PDAS, PERSONAL COMMUNICATORS, AND MOBILE COMPANIONS

8.1 INFORMATION APPLIANCES

The first wave of personal digital assistants (PDAs) and personal communicators were greeted with a barrage of criticism. But do not be fooled: handheld information appliances are destined to succeed. Today, business people are inundated with phone calls, faxes, and e-mail messages. Given the opportunity, they will gladly delegate routine telecommunication tasks to machines. Rather than employing PDAs and personal communicators to "communicate anytime, anywhere," users will rely on them to automatically screen, reroute, and even reply to calls and messages.

Wireless information appliances will become indispensable tools for knowledge workers—people who depend on timely access to information. A cottage industry of information hunters and gatherers will feed a steady diet of news and data to wireless broadcast networks. Information appliances will not simply comb the airwaves for keywords—they will be trained to recognize their user's interests and act accordingly. A consumer market will appear sooner than expected; business people are also consumers, and they are finding it increasingly difficult to keep their work and family lives separate. As employers demand longer hours, they will be forced to provide employees more effective means of staying in touch with family and friends, or risk possible backlash.

Information appliances trace their roots to pagers, calculators, and laptop computers. Vendors have different ideas about what these devices must do, which configurations are best, and what they should be called. Apple Computer's Newton®, although highly innovative, is modeled after the personal organizer. Other devices more closely resemble pagers and cell phones, and they go by an assortment of names: personal communicators, PDAs, personal information processors (PIP), personal information appliances (PIA), mobile companions, tablet computers, and intelligent cell phones—to mention just a few.

Some observers feel that the fledgling PDA industry promised too much too soon. John Sculley, then Apple Computer's chairman and CEO, vigorously promoted the PDA vision, and boasted his firm would soon transform itself into a consumer electronics powerhouse. For a handheld device to act like a (human) personal assistant, however, extraordinary handwriting and voice recognition, incredible speed and memory, and a powerful communications infrastructure are necessary. In retrospect, PDAs should have been positioned more modestly: as pocket fax machines, message centers, and data collection terminals. PDAs may not be ready for prime time, but they are exhibiting compelling benefits in health care, transportation, education, and retailing.

8.2 IS THE PEN MIGHTIER THAN THE KEYBOARD?

The first portable computers were actually transportable computers. They were easier to move from place to place than standard desktop machines, but they were not really suitable for use en route. Size, weight, and limited battery life discouraged users from lugging them through airports. The laptop succeeded primarily as a computer that could be shuttled between home and office.

Later, notebook, subnotebook, and palmtop computers capable of operating for several hours off batteries and small enough to fit inside a briefcase appeared on the scene. Still, these products targeted business travelers. By the late 1980s, vendors began to realize field workers would require a very different sort of machine. Unlike the personal computer, which was modeled after the typewriter, field workers needed a device that could replace the traditional clipboard and pencil. Equipped with a stylus for data input, the tablet computer promised to work just like paper—only better.

In 1989, Grid Systems introduced its GRiDPAD®. Later that same year, two companies focused exclusively on pen computing were formed: Go Corporation and Momenta Corporation The excitement over pen computing began to build. As Microsoft wrote in its backgrounder on Windows for Pen Computing™, ". . . continuing advances will make computers more personal, making them indispensable and something you reach for naturally whenever and wherever you need to retrieve, record or analyze information." Perhaps desktop microcomputers were mere precursors to the real personal computers.

Pen computers were supposed to combine the power of the microprocessor with paper's ease of use. Electronic forms would save time, ensure accuracy, and reduce data entry delays. For example, when a field sales representative needed to generate a price quote, an electronic form would automatically fill in prices and suggest possible options and accessories. Using mobile communications, an order could be zapped instantly to the factory, and the factory could respond with a firm delivery date. Unfortunately, pen computers did not take off as rapidly as expected; by 1994, both Momenta Corporation and Go Corporation were defunct.

But the industry has by no means given up on the pen as an alternative to the keyboard. In fact, companies such as Computer Intelligence Corporation (CIC) see applications for pen tablets not only in the field, but at the desktop and other fixed locations. Stylus input can be used in three ways. First, the pen may be used like a mouse for pointing, highlighting, and selecting. Second, the pen can be used to record *electronic ink*. In addition to drawing, electronic ink is useful for taking notes, sending quick fax messages, and recording signatures. UPS has deployed Inforite clipboard computers for this purpose. Pen computers can even verify signatures based on pen velocity, acceleration, and number of strokes—making forgeries all but impossible.

Much of the early enthusiasm surrounding pen computers was based on a third application: handwriting recognition. As the user enters characters or words, they are translated into their ASCII equivalents and redisplayed as computer text. Handwriting recognition also enables the application to detect incorrect entries and obviates the need to transcribe the data later.

The ability to read ordinary handwriting with 100% accuracy is no trivial task. (Consumers even imagined pen computers would be able to read handwriting that appeared illegible to them.) There are a number of challenges. For example, handwriting styles vary; people's handwriting tends to change when they are fatigued, and many people intermix printed and cursive letters. And of course, not everyone writes legibly.

Other factors degrade handwriting recognition performance. The display contains a digitizer that reports the pen's location as it moves along the screen's surface. Rather than smooth lines or curves, what the recognition software sees is a series of dots. While the human brain tends to filter out handwriting distortions, digitizers tend to amplify them. Distortion is primarily caused by parallax (in which the user perceives the pen and writing to be at different locations) and pen tip travel. Because the digitizer and display exist on different planes, parallax occurs, and the user may attempt to compensate by moving the pen (Figure 8.1). Pen tip travel arises when the digitizer cuts off the beginning of a pen stroke or adds a hook at the end; this happens because the digitizer responds to a combination of pressure and travel. Both types of distortion may cause characters to be recognized incorrectly.

There are a number of tradeoffs that can greatly improve the efficacy of handwriting recognition. According to ParaGraph International's president Stepan Pachikov, "In the near future, it [will be] possible to develop a handwriting recog-

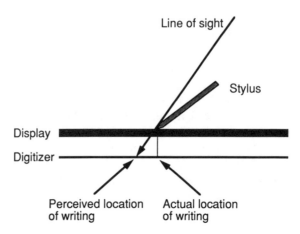

Figure 8.1 Handwriting distortion is caused when the user perceives the pen and writing to be at different locations due to parallax.

nizer that can recognize handwritten text as well as a human." (ParaGraph developed the handwriting recognizer used in the Newton.) But Pachikov points out this can only be accomplished by using context (e.g., word dictionaries) with the aid of additional processing power and memory.

Accuracy may be improved in other ways. The system can be trained to recognize the individual user's handwriting. The user can be forced to write each character in a particular place using onscreen boxes or *combs*. Accuracy may be improved by matching whole words to a dictionary (rather than attempting to recognize characters one at a time). Electronic forms can facilitate recognition by taking into account the purpose of specific fields; for example, the software need only look for numbers in a field labeled "Quantity."

Another—but highly controversial—way to improve accuracy is to teach the user a modified alphabet. Palm Computing's Graffiti® eliminates ambiguities due to capitalization, multiple strokes, similarities in shape, and the complexity of punctuation. The basic alphabet is reduced to a set of single-stroke characters the firm claims can be learned in just minutes. The use of single-stroke letters also speeds recognition—the recognizer does not have to wait to ensure what looks like an *F* will not turn into an *E*. Capitalized letters, numbers, and punctuation can only be entered after *shifting* (by tapping on the screen) into the appropriate mode. The Graffiti alphabet is shown in Figure 8.2.

In the future, PDAs may incorporate voice recognition technology. This would certainly make PDAs more humanlike. But it is still easier to recognize handwriting than voice: voice recognition systems require pauses between each spoken word, forcing the user to speak in an unnatural manner. It is also easier to edit pen input, and pen proponents point out that voice is simply inappropriate in certain situations, like taking notes during a meeting.

Figure 8.2 Palm Computing's Graffiti transforms the alphabet into a series of single-stroke characters for enhanced handwriting recognition accuracy.

8.3 PERSONAL DIGITAL ASSISTANTS

PDAs are pen computers with integrated personal information managers (PIM). PIMs are computerized address books, appointment calendars, and to-do lists. Apple Computer hoped that, through a combination of handwriting recognition and intelligence, its Newton PDA would supplant popular personal organizers made by companies such as Sharp and Casio. While the first Newtons may not have been a success, neither were the first Macintosh computers. When the industry ventured into this new category, many understood it would probably take several product iterations.

PDAs have been, however, more successful than is generally acknowledged. Approximately 175,000 devices based on the Newton operating system were sold in little over a year. While perhaps 75,000 units were grabbed up by industry insiders and gadget freaks, Newton has also gained a foothold in a number of major corporations. There are also over 1,000 software developers working on Newton applications. In addition, pen-based personal organizers have done quite well in Japan, where three different character sets (one containing thousands of characters) greatly complicate the use of keyboards. (One marketing theory is that sales of new high-tech products like Newton often enter a chasm after the initial flurry of sales to early adopters. During this period, the manufacturer must struggle to build sales in vertical markets where there are compelling applications—compelling enough that customers will be willing to pay higher prices, absorb risks, and learn how to use the product in unproven applications.)

There are two major PDA camps: Newton and Zoomer. Both systems have been licensed and are manufactured by multiple vendors. The original Newton was more powerful, but relied on real-time handwriting recognition. Zoomer was less expensive and consumed less battery power; it emphasized electronic ink and deferred recognition (performing handwriting recognition offline). Let's take a closer look at these two camps.

Apple's Newton MessagePad: If Macintosh was the desktop computer "for the rest of us," Newton was supposed to be so easy to use that it should not even be called a computer. Apple described Newton as an electronic notepad designed to assist users in "capturing, organizing and communicating ideas and information" and capable of recognizing both printed and cursive handwriting.

Newton integrates proprietary intelligence, recognition, information, hardware, and communications architectures. It can learn about the user's habits and preferences over time and propose solutions to help the user work more efficiently. Newton makes "simultaneous use of several recognition technologies," even handling freeform entries (characters not restricted to boxes or combs). Its object-oriented data structure allows users to access information in a variety of ways. (Object-oriented operating systems are composed of interchangeable modules, or "objects", which software developers can draw upon to perform common tasks.) Newton's reduced instruction set computer (RISC) processor was said to have the equivalent processing power of "leading desktop computers, yet consum[ing] less battery power than a small flashlight." And Apple claimed that Newton was being designed "from the ground up" to support communications. In summary, Newton was to be a "universal in box and out box" for sending and receiving memos and faxes.

When it was finally released in the summer of 1993—well behind schedule—the Newton MessagePad 100 fell far short of expectations. On the surface, everything was OK: the unit measured $1 \times 5 \times 7$ inches, weighed 15 ounces, and featured a single PCMCIA Type II slot. But all was not well behind the scenes; the struggle to bring the Newton to life reportedly led to a death—the suicide of Apple programmer Ko Isono (as described in the December 12, 1993 issue of the *New York Times*).

The Newton MessagePad 100 was greeted with a hailstorm of criticism. A number of journalists and industry analysts complained Newton's handwriting recognition did not work; the technology was even lampooned in the popular *Doonesbury* cartoon series. But the MessagePad's handwriting recognition was arguably better than any pen computer on the market, and it even improved over time as Newton learned the user's style. If a word is misinterpreted (using a built-in dictionary), an onscreen, pop-up keyboard may be summoned. In fact, the problem with the first Newton was not its handwriting recognition—it was the lack of an integrated dial-up or radio modem.

Newton did deliver some of the promised intelligence. The unit features a built-in notepad, name card file, and calendar with a to-do list. If you write "Lunch Joe Thursday" and tap the Assist icon, Newton will open its calendar and make a tentative appointment for lunch (12:00 to 1:00 p.m.) with the Joe in your name card file whom you often reference by first name. A window opens to confirm the appointment, but also provides quick access to the name card file in case you meant a different Joe.

The Newton MessagePad is based on a RISC processor made by Advanced RISC Machines: the 20-MHz ARM610. Newton features a built-in infrared port for beaming data to or from other MessagePads at distances of up to 1m (at 19,200 bps). Additional communication capabilities—such as a dial-up fax modem, paging receiver (Motorola MessageCard), and two-way radio modem (for RAM Mobile Data's Mobitex network)—must use the unit's PCMCIA card slot or serial port. Ex Machina's Notify! software enables the Newton to receive alphanumeric pages.

The original MessagePad offered insufficient random-access memory (RAM), could not store handwriting as electronic ink for deferred recognition, and used four AAA batteries that required frequent replacement. The first version of the Newton Connection Kit—for sharing data with desktop Macintosh computers—was not capable of exchanging data with popular database programs. Most of these problems, however, were corrected later (Figure 8.3).

But the Newton MessagePad has shaken the data collection terminal industry out of its slumber. Companies like Telxon (Akron, Ohio), Symbol Technologies

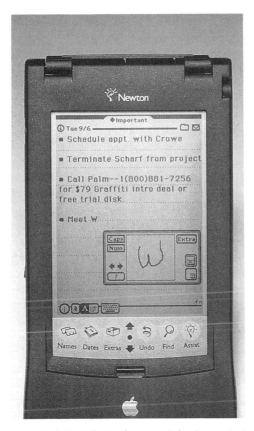

Figure 8.3 The Newton MessagePad 110 is depicted running Palm Computing's Graffiti; characters must be entered in a special box, as shown.

(Bohemia, New York), and Norand (Cedar Rapids, Iowa), who developed special-purpose handheld data collection terminals, now recognize the value of more flexible platforms like the Newton. These firms have begun to introduce PDAs of their own. For example, Symbol Technologies believes its PDA will complement the more specialized terminals used by stock clerks, giving store managers quick access to department and storewide information and salespersons the ability to check stock or enter orders. And it is in precisely these kinds of applications that Newton is slowly developing a dedicated following.

8.3.1 Zoomer

Tandy Corporation and Casio introduced Zoomer at the Consumer Electronics Show in June 1993. The pen-based Zoomer uses an icon-based operating system developed by GeoWorks (Berkeley, California). According to Casio's president John J. McDonald, consumers will be able to ". . . gain all the benefits of personal electronics without the typical steep learning curve required of PCs." Tandy launched its version at $699, with distribution through its nationwide chain of Radio Shack stores.

Like Newton, Zoomer is more of a personal organizer than a communicator. Zoomer's only built-in connectivity solution is its 9,600-bps short-range (2.5m) infrared port. The unit features a PCMCIA Type II card slot which can accept wireline and wireless communication peripherals. It also includes built-in software (in ROM) for the Vienna, Virginia–based America Online service. But Zoomer did have one clear advantage over the original Newton MessagePad: its ability to operate for up to 100 hours on three AA batteries (Figure 8.4).

Zoomer devices use the GEOS operating system and user interface. According to GeoWorks' CEO Brian Dougherty, the firm originally developed GEOS as a more compact (only 512K of RAM) and robust alternative to Microsoft Windows®. By the Fall of 1991, the firm realized it had been beaten by Microsoft's marketing machine. Dougherty recognized, however, that consumer-oriented, handheld devices (what GeoWorks calls *consumer computing devices* (CCD)) could turn out to be a big market for compact operating systems. In fact, Dougherty believes there are consumer markets for low-cost PDAs, and opportunities for telephones and televisions that are compatible. GeoWorks was able to sell Casio, Tandy, Sharp, AST, and Canon on the revised GEOS 2.0, which was designed to stretch the performance of low-cost, low-power 8086- and 8088-compatible microprocessors. Subsequently, the firm received additional funding from DOS-compatible palmtop manufacturer Hewlett-Packard, and LAN operating system giant Novell.

GEOS 2.0 is an object-oriented multithreaded, multitasking operating system. Because it is object-oriented, it facilitates smaller applications (typically under 100K) and shorter development cycles. GEOS 2.0 includes a rich text object, spreadsheet

Figure 8.4 The Casio Z-7000 Zoomer PDA: are consumers ready to buy it?

object, flat file database object, and graphics object library. Other built-in objects handle tedious chores like file management and text manipulation, relieving developers from reinventing similar functions in each new application. Multithreading breaks tasks into multiple pieces for simultaneous processing. GEOS 2.0's threads are prioritized, with the user interface assigned top priority. For example, if a complex object is being drawn and the user decides to turn the page, GEOS does not force the user to wait for the drawing to be completed as it would if it did not incorporate multithreading (Figure 8.5).

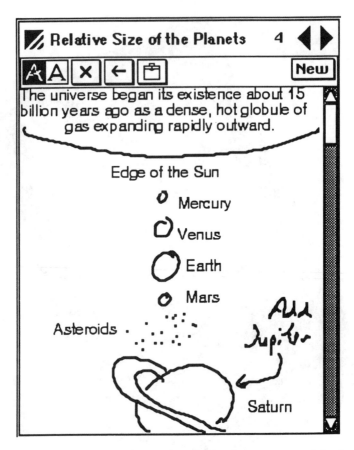

Figure 8.5 The GEOS user interface permits users to enter handwritten notes and sketches.

GEOS is now available on the Sharp PT-9000, Tandy Z-550, and Casio XL-7000 Zoomer class devices. The Z-550 includes notebook, calendar, to-do list, address book, alarm clock, calculator, and world time clock functions. It features a built-in dictionary (50,000 definitions), spell checker (100,000 words), thesaurus (660,000 synonyms), and 26 language translators for commonly used phrases. The Z-550 also includes world, U.S., and consumer travel information and comes with Pocket Quicken financial software for managing expenses and sharing related data with a desktop PC. Handwriting and text may be freely intermixed in notes, lists, sketches, and maps. The backspace function not only works with printed text, but with electronic ink by deleting individual pen strokes in the reverse order in which they were entered.

In many ways, Zoomer was more practical than Apple's first Newton. Its longer battery life, emphasis on electronic ink over handwriting recognition, and built-in software made it useful right out of the box. But Zoomer's architecture is inherently less powerful than Newton's. Zoomer manufacturers, software developers, and add-on hardware producers may have taken all of the talk about consumer markets too literally. While desktop PC packaging, advertising, and distribution have become more consumer-oriented, customers are still predominantly business users. At this early stage, it appears that Newton, with its greater processing power, will be the winner.

8.4 PERSONAL COMMUNICATORS AND INTELLIGENT AGENTS

Unlike PDAs, personal communicators are clearly pen-based communication devices. Three different camps have emerged: devices built around the dial-up phone line, the cell phone, and the packet radio. (For a comparison of PDAs, personal communicators, and mobile companions, see Table 8.1.)

In high-tech markets, being first is no guarantee of success. Often, other vendors learn from pioneers' mistakes. (But it is true that the most spectacular successes are enjoyed by pioneers who manage to stay on top.) AT&T-backed EO Corporation introduced the first personal communicators in 1992—the EO 440 and 880—but by 1994 the firm was out of business. Still, the experiences of EO and its partner Go Corporation serve as invaluable lessons toward understanding the personal communicator market.

Table 8.1

Comparison of PDAs, Personal Communicators, and Mobile Companions

	Apple Newton	*Casio Zoomer*	*BellSouth Simon*	*Sony Magic Link*	*Motorola Envoy*	*WinPad (various)*
Operating system	Newton	GEOS	Simon	Magic Cap	Magic Cap	Mobile Windows
Processor	ARM 610	8088	×86	Motorola 68349	Motorola 68349	Low-power ×86
Size (inches), weight	1.25 × 4 × 8, 1.28 lb	1 × 4.2 × 6.8; 1 lb	1.5 × 2.5 × 8; 1.1 lb	1.0 × 7.5 × 5.2; 1.2 lb	1.2 × 7.5 × 5.7; 1.7 lb	Unknown
Price	$599	$599	$899	$995	$1,500	Unknown
Infrared	√	√	No	√	√	Unknown
Other connectivity	PCMCIA pager; Metricom, Digital Ocean, ETE attachments	PCMCIA pager; RAM Mobile Data, cellular, add-ons	Cell phone, fax/data modem; PCMCIA pager	Data/fax send modem; PCMCIA pager	Data/fax send modem & ARDIS radiomodem; PCMCIA pager	Unknown

8.4.1 EO 440/880

When pen-computing pioneer Go Corporation was founded (by ex-employees of the software giant Lotus Corporation), it planned to develop hardware and software solutions. Later, the company decided to position its operating system as hardware-independent, so it spun out its hardware design team in 1991. That spinout became known as EO Corporation. By the middle of 1993, AT&T had acquired a majority interest in EO. (Interestingly, Go and EO were remerged later that same year.)

EO introduced the world's first personal communicators—manufactured by Japan's Matsushita—on 4 November 1992 (Figure 8.6). At the time, EO's president and CEO Alain Rossmann predicted that ". . . personal communicators will have as much impact on person-to-person communications as the telephone had in the early 1900s." And he was probably right.

These first products were the tablet-sized EO 440 and 880, built around AT&T Microelectronics' Hobbit processor and Go Corporation's PenPoint operating system. Nine applications were bundled with each model for functions such as faxing, note taking, sending/receiving e-mail (each was sold with an AT&T Easy-Link mailbox), scheduling, maintaining an address book, and calculating. The EO 880 even included a built-in microphone and speaker for recording and playing back digitized audio. But prices started at $2,000. Although the units featured a built-in 14,400-bps V.32bis/9,600 fax dial-up modem, the cell phone was optional and added $799 to the base price. Simply put, the EO personal communicators were too expensive for most business users.

Go Corporation's PenPoint (released April 1992) was designed around a notebook metaphor. It is a multitasking object-oriented operating system with a number of innovative features, including: (1) its notebook user interface (NUI), (2) embedded document architecture (EDA), (3) mobile connectivity support, and (4) a compact and scalable design. The NUI presents documents as pages within a notebook and includes a table of contents, selectable tabs (in the right-hand margin), and page numbers.

The EDA allows users to embed documents within documents—all of them controlled from a single master. For example, a letter may contain text and drawings, and both the word processor and drawing program remain "live" applications. The user can drag text into the drawing, or vice versa.

Mobile connectivity is facilitated by special features. While most desktop operating systems require tens of seconds after power is turned on before they are ready for use, PenPoint boots up almost instantly, automatically opening the last active application and document. Deferred input/output permits mobile users to work as if always connected to a network. For example, the user may prepare and send a fax document, but actual transmission may not take place until the unit is reconnected to a phone line or moves back within range of a radio network. Wireless communications is also facilitated by PenPoint's ability to suspend and resume communication sessions.

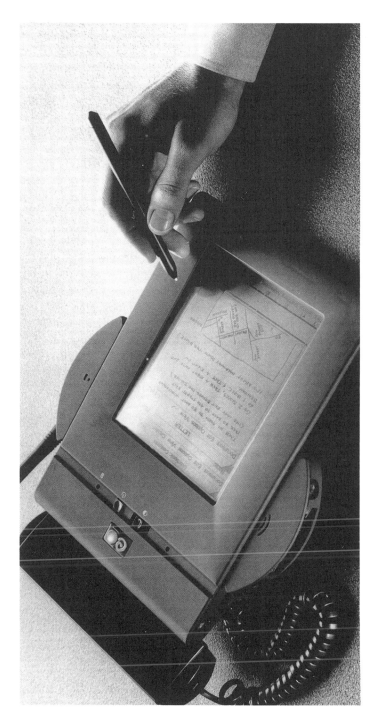

Figure 8.6 The EO 440 with optional cellular telephone was the first pen-based personal communicator.

Go eventually migrated from its pen-centric notebook metaphor to a more communications-centric envelope and fax cover sheet metaphor. The GO Message Center was introduced, consisting of GO Mail, AT&T Mail Communication Link, GO Address Book, Dialing Location Sheet (for automatic adaptation to the user's current country code and area code), and systemwide In Box and Out Box. GO Mail provides a single interface (fax cover sheet) for e-mail, faxing, paging, and other services. Functions made available to all applications included send, receive, reply, forward, and attach. Compatibility with various networks was provided through snap-in software "communication links."

While PenPoint was EO's operating system, AT&T's Hobbit microprocessor was its brain. Hobbit represents a clean break with industry-standard computer architectures (i.e., Intel and Motorola CPUs), since it was designed specifically for pen- and communications-centric devices. AT&T's goal was to develop chips that integrated voice, data, handwriting recognition, fax, electronic mail, still images, and ultimately full-motion video in one device. The company envisioned a future filled with Hobbit-based products tentatively called Tablet Communicators, PhonePads, Travel Companions, CellPads, NotePhones, PowerPhones, and Video FlipPhones.

The Hobbit is a C-Language Rational Instruction Set Processor (CRISP). It supports preemptive multitasking, integrated power management, and communications facilities. Hobbit's power efficiency is owed to fewer transistors, its complete 3 volt design, a resistorless local bus, and onchip caches to minimize external memory access. Combined with a digital signal processor (DSP) complex, Hobbit-based devices can integrate data compression, error control, modulation, and analog-to-digital and digital-to-analog conversion in one unit. A complete Hobbit chip set costs about $100.

With a platform based on Go Corporation's PenPoint operating system and AT&T Microelectronics' Hobbit, why did the EO 440 and 880 fail? First, EO took inflated market forecasts too seriously. Second, its products were too expensive—over $3,000 with the optional cell phone. Third, EO's products were introduced with a pen-centric rather than communications-centric user interface. Simon—the product we will consider next—appears to have corrected most of these problems.

8.4.2 Simon

BellSouth's Simon is the Swiss Army Knife of handheld cell phones. It combines fax machine, pager, personal organizer, and pen-based sketchpad/notetaker with a familiar cell phone. Designed and manufactured by IBM, the Simon Personal Communicator is marketed by BellSouth Cellular Corporation (Atlanta, Georgia) and was introduced at a suggested retail price of $899 (Figure 8.7).

Simon measures 8 × 2.5 × 1.5 inches and weighs 18 ounces. Its liquid crystal display (LCD) screen supports both pen and touch input. The unit features a multitasking operating system, Intel-compatible x86-series processor, 1 MB of RAM, 1

Figure 8.7 The traditional cell phone keypad is displayed on the screen of BellSouth's Simon.

MB of ROM, and a single PCMCIA slot. Its rechargeable NiCad battery delivers 8 hours of standby and 1 hour of talk time (at 0.6W of transmit power).

Simon looks and operates like a conventional cell phone—except the keypad is displayed on its screen. In addition to a pop-up QWERTY keyboard (i.e., standard typewriter configuration), Simon offers a unique "predictive keyboard" developed by IBM. The predictive keyboard displays six large keys representing the most likely next letter (with accuracy claimed at 90% to 95%). The predictive keyboard helps nontouch typists avoid hunting and pecking—improving data entry speed. (However, the predictive keyboard does have its drawbacks, since it is not very accurate when entering proper names, and fatigue sets in after only slightly extended use.) Simon's built-in applications include an appointment calendar, calculator, world clock with alarm, electronic directory, communications software for sending and receiving faxes, and remote access to LAN-based cc:Mail.

Simon uses electronic ink (what BellSouth calls *pen annotation*) rather than handwriting recognition and may be used as a notepad. Users can dial directly from Simon's electronic address book by simply tapping on the desired phone number. Simon also functions as a Group 3 fax machine using its built-in 2,400-bps Hayes-compatible cellular modem. Faxes may be either handwritten or typed.

The Simon team certainly did its homework. The unit is a fascinating (if not revolutionary) combination of existing devices: pager, cell phone, pen-based personal organizer, e-mail terminal, and fax machine. While various users may desire one or more of these items, it is not clear how many are willing to pay $900 for an all-in-one device. For example, many people own a cell phone and a pager; rather than carrying both, they may prefer to use the cell phone when commuting and the pager when on the road.

8.4.3 Magic Link

Sony's PIC-1000 Magic Link personal intelligent communicator is the first personal communicator based on General Magic's Magic Cap™ (Communicating Applications Platform) operating system and implementing the Telescript™ remote programming language. According to Sony, Magic Link ". . . simplifies and speeds . . . communicating, scheduling, organizing and coordinating." The unit is similar to Motorola's Envoy (see below), but is based on a different communications strategy. While the standard Envoy supports two-way mobile data, Magic Link uses an optional pager card for message notification and a built-in dial-up modem for responding. Both products support the new AT&T PersonaLink Service. Magic Link was introduced at a suggested retail price of just under $1,000 and comes bundled with financial management, spreadsheet, spell checking, and game software. The unit is shown in Figure 8.8.

Figure 8.8 Sony's Magic Link showing the desktop scene from General Magic's Magic Cap system software.

The user may operate Magic Link using a stylus or finger. The Magic Cap user interface provides a unique environment for keeping track of appointments, expenses, addresses, telephone numbers, telephone calls, billing preferences, and time zones. The "home" screen depicts a street with various buildings such as the user's home and office; additional buildings can be added by software developers. The user can go inside each building and each room. Items within rooms such as file cabinets, in and out boxes, desk drawers, and card files may be opened and used.

Magic Link employs a postcard metaphor for sending and receiving messages. In conjunction with AT&T's PersonaLink Service, postcards containing text, electronic ink, graphics, sound clips, and even animations may be exchanged. (Magic Cap includes an assortment of graphics, sound, and animation "stamps" that may be placed on the postcards.) When a message is received from someone not in its card file, the unit asks if a card should be created with the new name and address. Magic Link can also distribute a message to a group of users, automatically transmitting the message in the medium preferred by each recipient (e-mail, fax, or alphanumeric pager).

Magic Link's brain is Motorola's Dragon I chip set, consisting of the 68349 processor (32 bits, 16 MHz, 3.3V, and 4K cache) and the Astro ASIC designed specifically for Magic Cap. Astro handles all interface functions, including infrared, audio, LCD touch screen, PCMCIA slot, and dial-up modem. According to Motorola, Dragon I's power management is so fine that the unit switches to power-saving mode even between pen strokes.

Magic Link includes one PCMCIA Type II slot and an infrared port for communicating at 38.4 kbps (half-duplex) at up to 3m. The unit includes a built-in fax (9,600 bps send only) and data (2,400 bps) modem. Magic Link also includes a built-in microphone and speaker for recording voice messages up to 20 sec in length. A rechargeable lithium ion battery permits up to 10 hours of continuous use and does not suffer from the memory effect that plagues NiCad batteries. (NiCad batteries must be fully depleted before recharging; otherwise, they tend to lose their ability to accept a full charge.)

Sony's gamble is that users will prefer its communications strategy—the combination of a dial-up modem and pager card—over the more expensive approach taken by Envoy. In other words, rather than depleting their battery more rapidly, spending an additional $500 up front, and paying monthly mobile data service fees, users will be willing to wait until they have access to a phone line.

8.4.4 Envoy

Motorola's Envoy features a built-in two-way radio modem. The first version operates over ARDIS's nationwide wireless data network. Although somewhat larger and heavier than Magic Link, Envoy supports instant two-way communications. Envoy will also come bundled with Sony Electronic Publishing Company's Official

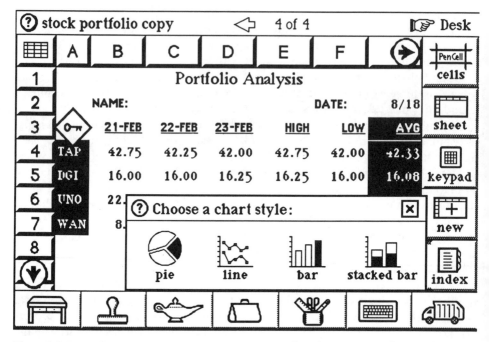

Figure 8.9 Personal communicator users may compose, send, and receive PenCell spreadsheets over AT&T's PersonaLink network.

Airlines Guide (OAG®) Flightline™ service. Flightline enables users to check available flights and make reservations while en route.

The PenCell™ spreadsheet and SmartWallet™ financial management software are bundled with both Envoy and Magic Link. PenCell enables users to send, receive, print, fax, or "beam" spreadsheets (or spreadsheet data) to desktop PCs or other Magic Cap–based personal communicators (Figure 8.9). SmartWallet assists users in keeping track (and later printing reports) of travel expenses. Perhaps one day personal communicator users will be able to purchase goods electronically: the buyer will beam credit card information to the POS terminal; the POS terminal will beam back a *virtual receipt*.

8.4.5 Magic Cap and Telescript

While Sony and Motorola possess world-class manufacturing capabilities, the real brains behind these devices is General Magic, which spun out of Apple Computer in 1990, taking with it some of Apple's best talent. Most notable are Bill Atkinson and Andy Hertzfield, inventors of the Macintosh personal computer. After nearly three years of secretive work, General Magic finally came out of the closet on February 8, 1993. The company's mission is to create personal intelligent communication products and services by developing and licensing technology to equipment manufacturers, network and information service providers, and software and entertainment companies. General Magic is no shoe-string operation, however. It has formed an alliance that includes Apple Computer, AT&T, Fujitsu, Matsushita, Motorola, Philips, Sony, and Toshiba.

General Magic hopes its products will play a role in everyday life, helping us remember (e.g., keep track of places to go, things to do), communicate (e.g., facilitate meetings, messages, and phone calls), and know (e.g., reservations, schedules, and directions). The firm believes its products will enable an entire new industry; but it will also require massive changes in our communications infrastructure. According to General Magic's chairman and CEO Marc Porat, implementing the company's plans may require two biologic generations.

Magic Cap is system software designed specifically for communications and capable of operating on both portable personal communicators and desktop PCs. All Magic Cap communicators use the Motorola Dragon processor and must conform to minimum memory, display, power management, and communication requirements. Magic Cap's user interface is an imaginary town populated with home, post office, office, and other buildings, and is intuitive and fun to use (Figures 8.10 and 8.11).

Figure 8.10 Magic Cap's top screen is a town inhabited by buildings representing its owner's home, office, and service providers.

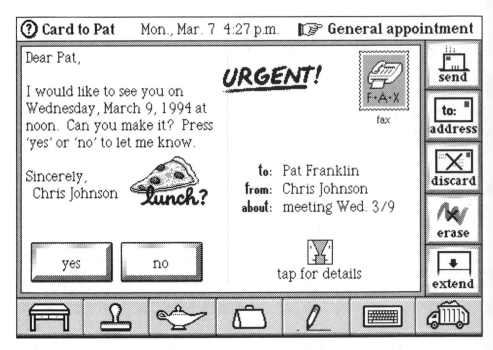

Figure 8.11 Magic Cap uses a postcard metaphor for creating and exchanging messages containing ASCII text, handwriting, graphics, sound bytes, and animations.

Telescript is what General Magic calls a *remote programming language*. Telescript's *smart agents* go places and act on behalf of their owners. In essence, they are messages that contain software programs. Security is managed by permits, authorities, and access controls. Telescript is object-oriented and will ". . . [embed] intelligence in electronic mail systems and computer applications, enabling them to cooperate with one another and with the networks that interconnect them." Telescript does not provide its own user interface, but enables RSVP buttons, smart envelopes, smart mailboxes (that can automatically filter and act on incoming mail), and the music, voice, graphics, and animation stamps mentioned previously. Magic Cap supports Telescript, but Telescript does not require Magic Cap.

The current communications paradigm—remote procedure calling—requires constant back-and-forth transmissions that force the user to remain on line for extended periods. In contrast, Telescript uses remote programming; messages are sent with executable code so that they may interact locally with the destination server. In the Telescript environment (designed by General Magic VP Jim White), user and merchant agents meet in the network (a place) and conduct business on behalf of their owners. Agents and places possess permits that determine how long they can live, who they can interact with, how much money they can spend, and other restrictions. Special authorities prevent agents from operating in dangerously anonymous ways. In essence, Telescript enables the establishment of an electronic marketplace.

It may not only take several years to implement General Magic's vision, it may take several years before we fully comprehend it. But one thing is becoming clear: personal communicators should not be evaluated in isolation. A new communications infrastructure—networks that can process rather than just switch messages, ubiquitous infrared access points, and embedded computers—will emerge to interact with these devices. Only then will we be able to fully judge their merit.

8.4.6 PersonaLink Services

AT&T PersonaLink Services aspires to become the town center for electronic merchants and their customers. Telescript-based smart agents—typically 500 to 1,000 bytes in length—will be transported to PersonaLink to find, buy, and sell things, filter information, and perform complex transactions. PersonaLink was placed in commercial service in September 1994.

PersonaLink is a new network with its own access points, network processors, and user mailboxes. Services include PersonaLink Mail and a daily news service from Mead Data Central (LEXIS®/NEXIS® online services). Wireless messaging—using one-way filtered rules-based delivery—will be supported via the SkyTel nationwide paging network. PersonaLink Market Square Services will provide a virtual three-dimensional shopping mall based on technology licensed from eShop™, Inc. Interconnection to AT&T's EasyLink electronic mail network will provide access to global e-mail and fax services.

PersonaLink is a radically new service. Today's messaging networks are message-switching networks; they simply store and forward messages. PersonaLink Services is a message-processing service; it acts on messages. For example, a user could send a message with the following special instructions attached: if the recipient does not retrieve this message from his mailbox by a certain date, convert the message to a fax and send it to the recipient's fax machine.

8.5 MOBILE COMPANIONS

Mobile companions represent a third camp: those who see handheld computers evolving from popular desktop PCs. Mobile companions will enable corporate planners to leverage their existing investment and familiar applications. These handheld devices will serve as portable file cabinets—giving users access to important text documents, spreadsheets, and databases while away from the office.

There is, however, another interpretation of the mobile companion strategy. Under the leadership of Chairman Bill Gates, Microsoft has done an outstanding job establishing the Windows™ operating environment as the desktop standard. Now the firm hopes to extend Windows to new platforms serving other markets. If Microsoft has its way, we will see Windows (or its derivatives) running on everything from handheld computers to TV set-top boxes.

8.5.1 WinPad

Microsoft first threw its hat in the mobile computing ring with Windows for Pen Computing—a set of pen-based extensions to the popular Windows graphical environment that support both existing and new applications. Given disappointing pen computer sales, however, Microsoft had to do more. The company is now working with third parties to develop handheld computers (code-named WinPad) based on its Microsoft at Work™ operating system. Microsoft at Work is the company's grand scheme for interconnecting Windows-based PCs with other appliances that inhabit offices and homes, such as fax machines, copiers, and TVs. WinPad devices will be able to access Microsoft Mail servers, synchronize data with desktop systems, and communicate over a variety of networks.

Companies developing WinPad products include Compaq Computer, Motorola, Toshiba, NEC, Olivetti, Sharp, and Zenith Data Systems. Chip makers Intel and VLSI Technology are developing low-power microprocessors compatible with Intel's industry-dominant architecture. WinPad will use Windows Open Services Architecture (WOSA) for remote communications and intermittent network access.

But it is clear that the first WinPad mobile companions are late—offsetting any advantage that would have accrued thanks to compatibility with desktop standards. In the end, the mobile environment will demand a clean break from the desktop. We saw it happen once before—when a small company named Microsoft sprinted past entrenched software vendors by developing an operating system for emerging microcomputers.

8.5.2 DTR-1

Dauphin Technology was founded in 1988 and scored a major coup when it won the largest portable computer contract in U.S. history: Lapheld II. The firm later made a splash with the introduction of its DTR-1 Desktop Replacement. The 486-based handheld computer accepts pen input, includes a minikeyboard, and supports multiple communications options (2,400/9,600-bps data/fax modem, Ethernet adapter, wireless LAN adapter, and radio modem for use over RAM Mobile Data's Mobitex network).

According to Dauphin's president Alan Yong, "The desktop computer, as we know it, is running on borrowed time." Yong sees handheld computers serving as companions to desktop PCs in vertical markets such as health care and insurance. Gradually, users will come to rely on handheld computers in place of desktop units.

But in order for this to occur, the pen must also be integrated with desktop computers.

8.5.3 CruisePAD

Zenith Data Systems (Buffalo Grove, Illinois) takes the concept of mobile companions quite literally. The firm's CruisePAD acts as a remote interface to the desktop PC. Using a wireless LAN developed by Proxim, CruisePAD enables remote operation of a desktop PC's display, keyboard, mouse, and audio. The mobile user can maintain access throughout a large building or campus thanks to Proxim's multicellular architecture. The CruisePAD approach enables portable users to grab text documents, spreadsheets, or presentations off their desktop PC, enter data in a database, or run programs remotely. There will no doubt be a large market for this type of device once it can be used over nationwide mobile data networks.

8.6 A KILLER APPLICATION?

Most personal organizers, palmtop computers, PDAs, and personal communicators come equipped with built-in infrared ports. While omnidirectional infrared LANs are expensive—signals must survive bouncing off walls, floors, and ceilings—users who are willing to point and shoot can have short-range infrared communications for a mere $1.50 to $4.50 in parts.

There are two major infrared standards: IrDA and Magic Beam. The IrDA serial infrared (SIR) standard is based on a proposal from Hewlett-Packard, and is inexpensive and simple to implement (see Chapter 7). Magic Beam is supported by General Magic, Sony, and Motorola and offers higher performance for the future. Both systems permit spontaneous device-to-device communications.

After several years, the wireless industry may have finally found a market for replacing cables. Ironically, it will be the least expensive cables that will be replaced first: short jumper cables. Point-and-shoot infrared will be used to link portable computers to desktop PCs, LANs, servers, and each other.

But the big prize could be walkup connectivity to pay phones. Imagine arriving at an airport and walking up to the nearest available pay phone. Instead of lifting the receiver and dialing from the keypad, you point your personal information appliance (PDA, personal communicator, or mobile companion) at the infrared port on the front of the phone. Then you tap on the "office" icon that appears on the screen of your device. Using an infrared link, your personal information appliance dials the phone and automatically charges the call to your credit card.

You are now connected to an intelligent voice mail system back at your office. It handles not only voice calls, but also e-mail and fax. Instead of talking to your receptionist, a list of messages appears on the screen of your personal information appliance. Each listing includes the phone number, fax number, or e-mail address of the originator; your handheld device searches its name card file and displays any matches.

You can respond to messages in a number of ways. First, you can download and review selected voice mail or e-mail messages. Faxes can be forwarded to your hotel, or faxes that are one page or less can be downloaded directly to your personal information appliance. Next, you can return important calls by merely tapping on the "return" button that appears after each voice mail listing. If you reach someone who would like one of your electronic business cards, you download it as a data or fax burst during or at the end of the conversation. You can also send and receive product information and even complete catalogs in this manner.

Of course, this application requires some extra support at both ends. At your end, you will need a pay phone with an infrared port, built-in fax/data modem, and the ability to switch between voice, data, and fax modes. At the other end, there must also be the ability to switch between modes; this will most likely be provided through a standard PC-to-telephone interface.

Radish Communications Systems' VoiceView® defines a mechanism for freely switching between voice, data, and fax modes under software control. The firm has also developed hardware products for adding display screens to phone sets and for linking telephones with nearby personal computers. VoiceView is implemented as a set of dial-up modem AT command extensions. According to Radish, VoiceView requires only a small change to existing modem technology. Radish envisions PC/phone integration for a number of applications. For example, users could send a fax by simply dragging the document on their PC screen over to a "telephone" icon. Nevertheless, the company faces a classic chicken-and-egg dilemma: VoiceView will not become fully accepted until there is a critical mass of users.

This capability would surely open the door to a large horizontal market serving business travelers. Ubiquitous infrared access points would relieve users from having to purchase expensive power-consuming and unreliable radio modems for communicating wirelessly from within buildings. And there would be no airtime charges—just standard phone charges.

8.7 CONSUMER MARKETS AND ELECTRONIC SHOPPING MALLS

Developers hope to create a consumer market for PDAs and personal communicators. While it is unclear whether they will succeed, one thing is certain: computer marketing will never be the same. In an attempt to penetrate consumer markets, computer manufacturers have adopted consumer marketing tactics. These techniques are proving effective because (1) they are generally more sophisticated, and (2) business users also happen to be consumers.

In the industrial marketing model, product development cycles are long (often taking two to three years) and considerable time elapses between product introductions. All of the market research must be accomplished up front. In the consumer marketing model, product development cycles are short (down to six months or less) and new products are frequently introduced. Most of the market research is conducted in real time: vendors develop a product, throw it on the market, listen to customers' reactions, modify the product, and begin the process again.

Why do some computer vendors prefer the consumer marketing model? They realize that no one can know for certain how radical new products will succeed. Customers can tell you what they like and do not like or what their current problems are, but they are not market planners. And they know even less about technology in the pipeline. The consumer marketing model permits the computer company to take an entrepreneur's dream and refine it until it either becomes a successful product or must be abandoned.

While we may not know how personal information appliances will succeed, there is plenty of evidence that they will succeed. Tremendous financial and human resources are being applied. Companies are developing highly creative and clearly differentiated products. New firms are being founded. With all of this talent and energy at play, someone is bound to stumble on a winning solution.

In the future, PDAs and personal communicators may not merely change the way we conduct business. *They will become the focus of all business activity.* Smart agents will spend digital cash in electronic shopping malls—with advantages accruing to both buyer and merchant. Buyers will always get the best deal; merchants will be able to reach all of the customers and at the lowest cost. Electronic shopping malls will not replace the free market; they will make it truly ubiquitous—screaming past walls erected by government regulators and central planners at the speed of light.

SELECT BIBLIOGRAPHY

[1]   Brodsky, I., "Coming! Information Appliances," *Wireless: For the Corporate User*, November/December 1992.

[2]   Cummings, M., "The Advance Guard in Hand-Held Computing," *Business Communications Review*, February 1993.

[3]   Dao, J., "Handwriting Recognition Systems: Design Issues and Philosophy," Communications Intelligence Corporation technology white paper, 1992.

[4]   Infrared Data Association, press kit, March 1994.

[5]   Markoff, J., "Marketer's Dream, Engineer's Nightmare," *The New York Times*, December 12, 1993.

[6]   Pachikov, S., "Handwriting Recognition for PDAs," presentation at Wireless Data'94, San Francisco, California, October 1994.

[7]   Scavuzzo, R., AT&T Microelectronics, speech delivered August 12, 1992.

[8]   Wagoner, J., "PDA Roundup," *Personal Electronics News*, November/December 1993.

CHAPTER 9

▼▼▼

MOBILE SATELLITE SERVICE

9.1 A SATELLITE TRANSCEIVER IN YOUR POCKET?

When most people think of satellite communications they envision backyard dish antennas or huge, sophisticated earth stations. A new variety of satellite communications, however, is furnishing wide-area voice and data communications to users on ships, planes, and trucks: mobile satellite service (MSS). But the next-generation MSS promises a great deal more: services that are truly ubiquitous, economical, and, in some cases, accessible to pedestrians equipped with handheld transceivers.

We owe this revolution in satellite communications to two major developments: satellite antennas that can focus signals into narrow beams—projecting *macrocells* onto the Earth's surface—and fleets of satellites placed in low-altitude orbits where they act like repeater stations on extremely tall towers. Although several hundreds of kilometers above our heads, satellites in low earth orbit may be easier to reach than base stations just a few miles away on the ground, since terrestrial signals must journey over and around uneven terrain and man-made obstacles.

Were it not for buildings, MSS would enable us to create truly ubiquitous mobile communications services. With satellites, there is little difficulty in extending coverage to the remotest of locations. For those who demand the ability to use one device in any location, MSS offers common air interfaces with continentwide and even global scope. But coverage is obscured on the streets of large cities, where tall buildings create a canyon effect, and within buildings possessing steel structures (except from directly behind nonleaded windows).

These limitations notwithstanding, industry is planning to pour billions of dollars into MSS networks. Most pledge to complement, rather than compete with, terrestrial-based services. Some are positioning themselves to provide cellular and PCS fill-in service for rural and remote areas. Motorola hopes its brainchild, Iridium (Washington, D.C.), will blossom into a worldwide cellular telephone and paging network, but assures terrestrial cellular operators that its high cost structure (and plans to develop dual-mode satellite/terrestrial phones) will discourage direct competition between satellite- and terrestrial-based systems. A small group of visionaries believe they can establish entirely new businesses such as search and rescue.

While the rest of the world was infatuated with high-altitude geosynchronous-earth-orbit satellite (GEOS) systems, a gaggle of U.S. technology innovators and business leaders developed plans for new services built around low-earth orbit satellites (LEOS) networks. Trading the wide-area coverage of a single GEOS for the improved accessibility (to subscribers) of a fleet of satellites, LEOS proponents hope to make satellites a major player in the personal communications revolution. In late 1993, the FCC allocated 33 MHz to LEOS: 1,610 to 1,626.5 MHz and 2,483.5 to 2,500 MHz. Other LEOS players, such as Orbcomm, Spaceway, and Teledesic, plan to use frequencies below 1 GHz and above 20 GHz.

Telecommunication satellites changed the world forever. We sometimes forget that a little over 30 years ago—amid much fanfare—satellites opened the era of real-time intercontinental television reporting and broadcasting. Now MSS promises the first worldwide mobile communications services. Ironically, the biggest market for MSS may have nothing to do with mobile communications, but with providing modern telecommunications services in remote areas of underdeveloped countries.

Nevertheless, proposed MSS networks carry considerable risk. High startup costs—ranging from hundreds of millions to several billion dollars per network—mean MSS networks must acquire hundreds of thousands and even millions of subscribers just to break even. (For example, Iridium hopes to acquire at least 1.5 million subscribers.) The belief that MSS will appeal to a significant fraction of existing cellular telephone users may be unrealistic. The best application fit would appear to be international business travelers. Time zone differences, however, may reduce interest in real-time global communications.

The biggest challenges facing MSS are right here on the ground. MSS carriers must raise huge sums of money. They must see to it that small, low-cost subscriber terminals are developed. But the biggest obstacle may be political: many countries fear MSS will be used to bypass domestic telecommunications services. In developing countries, government often owns and operates the public telephone network, and international phone calls placed by foreign visitors are a lucrative business. MSS players will also have to go through the arduous process of obtaining licenses for each country in which they wish to offer service; but they will be forced to turn off their system as it flies over each country that refuses to issue a license.

9.2 A BRIEF HISTORY OF SATELLITE COMMUNICATIONS

In 1945, Arthur C. Clarke authored the first article suggesting artificial satellites could be used for telecommunication purposes. Thirteen years later, Bell Labs proposed bouncing voice, music, and data signals off a satellite. On 12 August 1960, the National Aeronautics and Space Administration (NASA) successfully launched ECHO 1 into a circular orbit and, along with Bell Labs, used it to relay signals across the United States. This success led to the development of the Telstar satellite, which was launched in 1962. That same year, the Communications Satellite Act was passed by the U.S. Congress.

The Communications Satellite Act established a quasiprivate company called the Communications Satellite Corporation (Comsat). This corporation was half owned by the public and half owned by common carriers including AT&T, Western Union, and RCA. Comsat was granted exclusive rights to establish international satellite communications for the United States.

The U.S.'s willingness to share its space technology for peaceful purposes led to the formation of the International Telecommunications Satellite Organization (Intelsat) global cooperative in 1964. Intelsat's 121 member nations jointly own and operate a 17-satellite global network used by 180 countries. Comsat is the U.S. signatory to, and owns approximately 25% of, Intelsat. As the U.S. signatory, Comsat is responsible for the daily business and operations of Intelsat within the United States.

The International Maritime Satellite Organization (Inmarsat) was established under the Maritime Satellite Act of 1978 to provide telecommunication services for ships at sea. Inmarsat boasts 64 member countries; the majority of financial support is provided by the United States, Japan, United Kingdom, and Norway. Comsat is also the U.S. signatory to, and owns approximately 28% of, Inmarsat.

For years, these two organizations dominated satellite communications, with Intelsat controlling the market for fixed services and Inmarsat cornering the maritime mobile services market. In addition, Comsat has spearheaded efforts to expand Inmarsat services to encompass small-boat, aeronautical, and land mobile users. Although Inmarsat's satellites are parked in geosynchronous orbits over the Indian, Atlantic, and Pacific Oceans, a large portion of the demand for land mobile communication services is found in coastal areas. Initially, service to land mobile users was provided through excess maritime capacity; Inmarsat is dedicating capacity for land mobile use on new satellites.

Steady growth in satellite deployment continued until 1986, when the space shuttle disaster and other launch vehicle accidents brought space launches to a screeching halt. While NASA and Europe's Arianespace—both enjoying government protection from competition—have dominated the space launch business, Russia's Khrunichev Enterprise and China's Great Wall Industry Corporation now offer lower cost service. A number of other firms provide rockets, including McDonnell

Douglas (Delta), Martin Marietta (Titan), General Dynamics (Atlas Centaur), and Space Services (Conestoga).

In the United States, the FCC grants satellite communication licenses and ensures compliance with ITU frequency and orbit assignments. In most countries, satellite communications is controlled by the government PTT authority, which may restrict or even prohibit the direct transmission of information to the outside world.

9.3 MOBILE SATELLITE TECHNOLOGIES

Telecommunication satellites are essentially repeaters on tall towers. While their range is superior to terrestrial repeaters, it is achieved at much greater cost. A satellite, therefore, is best suited to serving a large population of users scattered over a wide geographic area.

Advances in space segment (satellites) and mobile terminal technology have made MSS possible. Sophisticated satellite antennas project patterns of cells onto the earth's surface, enabling frequency reuse. Satellites in low earth orbit can be reached by low-power handheld terminals. Digital multiplexing enables hundreds of users to simultaneously access a single satellite transponder.

Geosynchronous satellites (GEOS) are deployed 22,000 miles above the earth. At that altitude, their orbital velocity precisely matches the earth's rotation, so the satellite always appears to be in a fixed position relative to the earth's surface. The advantages of GEOS are that they can cover entire continents, and antennas on the earth always know where to find the bird in the sky. One GEOS can provide coverage over an area as large as North America 24 hours per day, 7 days per week.

But there are also disadvantages to operating at such high altitudes. First, placing and maintaining a satellite in a geosynchronous orbit is very expensive. An additional booster is required to reach 22,000 miles. The orbit must be carefully maintained; otherwise, the satellite will wander from its desired location.

Second, the round trip (up-down and down-up) is 88,000 miles, representing nearly one-half second of propagation delay (more than 10 times the delay one experiences during a terrestrial coast-to-coast telephone call in the United States). A half-second delay renders two-way voice conversation awkward, particularly when one party tries to interrupt the other or simply misinterprets a pause. In data applications, protocols may need to be modified to tolerate long delays, especially for functions such as receiving acknowledgments and poll responses. Third, due to sheer distance, there is greater signal attenuation, which is partially offset by improved antenna and receiver designs.

High-elliptical-orbit satellites (HEOS) and LEOSs are deployed at lower altitudes and consequently do not remain in fixed positions relative to the earth's surface. To provide continuous coverage to points on the earth's surface, HEOS and LEOS must be deployed in fleets in carefully spaced orbits so that as one satellite disappears over the horizon another will rise to take its place.

LEOSs have many advantages. While one GEOS may only serve a single continent, a LEOS fleet typically serves all or most of the world. Thanks to their low altitude, LEOSs are generally accessible to low-power mobile terminals with simple whip antennas. In fact, communication between LEOS and mobile users is possible at UHF and even VHF frequencies, enabling the use of low-cost, mass-produced components. The deployment of a LEOS fleet is less expensive and risky; multiple satellites may be launched via a single rocket. And compared to a GEOS, the propagation delay is usually much smaller.

Both "little LEOSs" and "big LEOSs" are planned. Little LEOSs operate below 1 GHz and are primarily targeted at low-speed messaging applications. Big LEOSs operate above 1 GHz, offer voice as well as medium-speed data services, and tend to be more expensive than their little LEOS cousins. The FCC is expected to license multiple LEOSs. (Three big LEOS players using CDMA technology have agreed to share the same frequencies, making the FCC's job easier.)

Satellites operate in five major frequency bands (Table 9.1). A mobile satellite link involves four channels: a mobile uplink, a mobile downlink, an earth station uplink, and an earth station downlink. The earth station uplink and downlink is called the *feeder* because it interconnects the satellite network with terrestrial networks. Earth station feeders generally operate on the higher frequencies where the attenuation is greater, because as fixed stations they can accommodate larger antennas and higher power transmitters.

The most popular satellite frequencies are found in the C-band, used by VSAT networks to interconnect widely scattered fixed locations. Several MSS players target the L-band for the mobile uplink/downlink to reduce costs. Others target the Ka-band because they wish to offer higher bandwidth services. Earth stations often target the Ku-band because it offers a combination of high bandwidth, small antenna size, and minimal interference (the Ku-band is free of terrestrial microwave users).

The useful life span of a satellite ranges from 5 years (for a LEOS) up to 15 years (for a GEOS)—primarily due to the limited ability of onboard propulsion systems to stave off orbital decay. The power system generally consists of both solar

Table 9.1
Satellite Communication Above 1 GHz Takes
Place in Five Major Frequency Bands

Band	Frequency (GHz)
L	1.2 to 1.6
S	2 to 4
C	4 to 6
Ku	11 to 17
Ka	20 to 30

cells and rechargeable batteries. Solar panels are configured to always face the sun. In outer space, however, solar cells are constantly bombarded by high-energy particles that gradually degrade their performance. A rechargeable battery provides power during solar eclipses, which for GEOSs occur just 1 hour per day over two 45-day periods each year.

The standard measure of a satellite's transmit signal is its effective isotropic radiated power (EIRP). EIRP takes antenna gain into account; it is the power an isotropic (omnidirectional) antenna would require to achieve the same results. For example, if a 40-dB gain directional antenna is used with a 1W transmitter, the EIRP is 10,000W. Maximum satellite EIRP has increased over time thanks to improvements in both power systems and antenna designs.

Up until recently, LEOSs have always targeted mobile applications. A joint venture between Microsoft's Bill Gates and McCaw Cellular's Craig McCaw hopes to provide global interactive multimedia communications to fixed users via a fleet of 840 LEOSs. (General Motors' Hughes division has proposed a 9- to 17-satellite GEOS network, called Spaceway, to do much the same thing.)

Teledesic would upgrade the entire planet's telecommunications infrastructure in one fell swoop. Even at several times its estimated cost of $9 billion (a figure critics contend is way low), Teledesic would be a fantastic bargain, offering services ranging from 64 kbps to 2 Mbps using the Ka-band. Although the project is still in the preliminary study phase, its backers hope it will become operational by the year 2001.

No doubt to avoid raising the ire of LECs and cable TV operators—and to win sympathy within the Clinton administration—Teledesic was proposed as an on-ramp for rural and remote users to the "information superhighway." Clearly, the market for broadband services is found in the major urban areas, not rural Kansas or the remote northern territories of Canada. Once off the ground, Teledesic could provide LECs and cable TV operators much needed competition in multimedia networking services.

9.4 APPLICATIONS

As submarine fiber-optic cable is laid between continents (AT&T has its own fleet of five cable-laying ships) and coaxial and fiber-optic cable is widely deployed on land, satellites no longer hold an exclusive grip on high-bandwidth, wide-area telecommunications. Not surprisingly, a sizable portion of the satellite industry is turning its attention to mobile communication growth opportunities.

MSS was introduced in the early 1980s. A modest-sized market has emerged; Comsat claimed approximately 18,000 ships and land mobile terminals were using Inmarsat by 1993. The maritime shipping industry has used MSS direct-dial telephony, fax, and data communications for over a decade. Real-time communication is used to monitor the condition and location of cargo (particularly hazardous materials), redirect ships, arrange port services, and receive weather data.

MSS is also well established in long-haul truck fleet management. Real-time data communication reduces out-of-route miles, engine idling, improves customer service, and provides emergency communications. This market is dominated by Qualcomm's OmniTRACS service, which has survived forays by SMR and meteor burst players. A 100-vehicle fleet can be outfitted with MSS terminals for roughly $4,000 per truck up front and $1,200 per year in usage charges, assuming each truck sends an average of six messages per day.

(Because it requires no external infrastructure, meteor burst has enjoyed success as a "survivalist" communications system, particularly in military applications. Meteor burst systems bounce signals off ionized gas trails left in the wake of the dust-sized meteors that continuously bombard the earth's atmosphere. Although the waiting time between sporadic meteor events is typically 3 to 5 min, meteor burst supports average throughputs of over 300 bps in the 30- to 50-MHz band. While communication is normally limited to 2,000 km, it is possible to have continentwide mobile data using repeaters. Meteor Communications Corporation (Kent, Washington)—the only long-standing commercial meteor burst player—has established a Fleet Telecom division to pursue the long-haul truck fleet management market.)

There is more competition on the way for OmniTRACS. TRX TransTel plans to construct a nationwide voice/data Telepoint network with base stations strategically located at truck stops, rest areas, and truck terminals. The firm believes it can succeed in the truck fleet management market by undercutting MSS prices, while offering better coverage than cellular telephone.

The commitment that long-haul trucking firms are willing to make to MSS is illustrated by a multimillion dollar agreement between Qualcomm and C. R. England and Sons (Salt Lake City, Utah). This refrigerated motor carrier is equipping 1,150 trucks with Qualcomm's OmniTRACS system. Drivers will be able to communicate directly with dispatchers via a keyboard/display unit mounted inside the cab, receiving updated itineraries and other time-critical information. The Omni-TRACS system will also enable dispatchers to monitor vehicle performance, trailer status (e.g., whether the refrigeration functioning properly), and other data in real time.

MSS offers advantages for commercial and private aircraft—particularly for transoceanic flights that take aircraft outside the range of terrestrial-based radar and communications systems. But perhaps the most lucrative opportunity is providing communication services for the more than one billion passengers who fly each year. Inmarsat offers an aeronautical telephone service called Skyphone—but at a whopping $10 per minute. On domestic flights within the United States, MSS must compete with existing air-to-ground telephone (ATG) services.

Providing fill-in service for cellular telephone carriers is also thought to be a major opportunity. Fifteen cellular operators have signed agreements with American Mobile Satellite Corporation to explore the possibility of becoming authorized MSS resellers. Celsat and Ellipsat are also pursuing cellular fill-in service opportunities. But dual-mode satellite/terrestrial phones will be quite expensive.

MSS could provide more than just fill-in coverage for PCS. Because PCS networks will have to be built from scratch, we can expect a prolonged period during which terrestrial systems will only deliver spotty coverage. A compatible MSS could provide near-ubiquitous service, and subscriber terminals would default to terrestrial PCS when within range. In addition to activity near the 1.8-GHz band, NASA's Advanced Communications Technology Satellite (ACTS) program is conducting experiments in the Ka-band to evaluate the practicality of handheld MSS terminals.

Will MSS become a major component of the virtual office? Given the complexities of integrating MSS transceivers with portable computers and the inability to serve users inside buildings, that seems unlikely. But there will be niche paging, wireless electronic mail, and fax applications for users who frequently travel to underdeveloped countries or other remote areas.

9.5 PROFILES OF SELECTED MSS PIONEERS

There are three geosynchronous MSS providers and several proposed LEOSs. Let's take a closer look at some of them.

9.5.1 American Mobile Satellite Corporation

In 1989, when the FCC evaluated license applications for the MSS, it was unable to identify a clear winner. The FCC recommended that the applicants form a consortium, to which it would issue a single license. Hughes Communications Mobile Satellite (29% owner), McCaw Communications (29%), and Mtel Satellite Service Corporation (29%) joined forces to establish American Mobile Satellite Corporation (AMSC) (Washington, D.C.).

Along with TMI Communications of Canada, AMSC has developed the North American MSAT (mobile satellite) system for mobile voice and data communications. Services will include circuit-switched voice, circuit-switched data, and packet-switched data. Prior to the release of commercial MSAT service (scheduled for early 1995), the two partners offered low-speed data services via the Inmarsat C satellite.

MSAT 1 and MSAT 2 will be launched in late 1994. The two satellites will serve the entire United States plus 200 miles off each coast. In conjunction with international partners, AMSC will offer seamless service between the United States, Canada, and Mexico. NASA has offered free launchings in exchange for 15% of the two satellites' capacity over a 2-year period. The satellites are being constructed by a team consisting of Hughes Aerospace Company (the launch vehicle) and Spar Aerospace Ltd. (communications payload) (Figure 9.1).

Each satellite is designed to operate for 15 years and will be equipped with an onboard propulsion system to maintain geosynchronous orbit. A 28-cell nickel hy-

Figure 9.1 AMSC's satellite will be placed in a geosynchronous orbit 22,000 miles above the earth's surface.

drogen battery will serve as backup to the solar power array (for when the satellite is in eclipse) to provide the necessary 3.15 kW of power.

AMSC claims it will provide toll-quality voice and 4,800-bps data services. Each satellite will support 4,200 channels that are 5 kHz wide. The mobile downlink will operate from 1,530 to 1,559 MHz, and the mobile uplink from 1,631.5 to 1,660.5 MHz. The earth station feeders will operate in the 10-GHz (downlink) and 13-GHz (uplink) bands. Circuit-switched service will cost $25 per month, plus $1.45 per minute of use. Packet-switched data service will be competitive with wireless terrestrial services, priced at about $0.15 per 500 characters.

AMSC is targeting two primary markets: cellular telephone and aeronautical radio. Services will include MSAT Telephone, MSAT FlightNet, and MSAT Private-Net. MSAT Telephone is a family of circuit-switched voice and data services linked to the public telephone network. It consists of Cellular Roaming (a cellular fill-in service), FlightCall (serving commercial airline passengers), and RuralServe (for rural homes lacking local telephone service). MSAT FlightNet will provide packet-switched safety-related services for the Federal Aviation Administration (FAA). MSAT PrivateNet will offer customized leased channel services to business users.

Mobile subscriber terminals are expected to cost $1,200 and up. Manufacturers such as Mitsubishi have announced plans to develop dual-mode (cellular + MSS) phones compatible with AMSC's service. Companies such as Rockwell will offer MSS terminals with integral GPS radiolocation receivers (Figure 9.2).

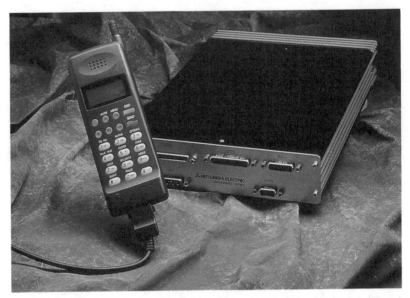

Figure 9.2 A combination vehicular MSAT terminal and handheld cellular telephone is being developed by Mitsubishi.

9.5.2 Celsat

This San Diego–based affiliate of Titan Corporation has proposed a hybrid personal communications network (HPCN) combining GEOS and terrestrial microcellular telephone networks. The company claims it will be able to cover the entire United States (with full roaming at no additional charge), charge less than $0.25 per minute of use, and serve users equipped with 0.1W handheld transceivers.

The key to Celsat's approach is a satellite antenna 20m in diameter that projects more than 100 spot beams onto the earth's surface—creating what Celsat calls *super-cells*. The system will use CDMA to support transmissions at speeds of up to 144 kbps. The firm claims it can offer compressed video, fax, and low-speed voice and data services.

9.5.3 Comsat

Comsat (Washington, D.C.) is the U.S. signatory to Inmarsat. Headquartered in London, Inmarsat offers a variety of services over L-band frequencies allocated on a worldwide basis. Standard A terminals operate over analog channel pairs on 1,636.5 to 1,645.0 MHz (4,800-bps uplink) and 1,535.0 to 1,543.5 MHz (1,200-bps downlink). There are over 10,000 Standard A terminals currently in use, primarily in voice applications. Due to their 3-foot-wide parabolic dish antennas and bulky transceivers, Standard A terminals are only marginally portable.

Standard C terminals operate on uplink frequencies between 1,631.5 and 1,646.5 MHz and downlink frequencies between 1,530.0 and 1,545.0 MHz at 600 bps in each direction. Standard C terminals are used primarily in messaging applications.

The newer Standard M supports smaller, more portable terminals operating at 4.8 kbps (voice or data). Standard B terminals are intended as the digital replacements for Standard A and run at 16 kbps. Comsat's High-Speed Data (HSD) service is a one-way 56-kbps service featuring a 3-kHz return channel, and may be used in conjunction with switched 56-kbps services over the public telephone network.

Inmarsat signatories have been investigating future mobile communications requirements in an effort called Project 21. Proposed Inmarsat P terminals would be dual-mode handheld devices for both PCS and MSS networks. In September 1994, Inmarsat announced the formation of a new company to build a $2.6 billion network consisting of 10 satellites (and two spares) in high elliptical earth orbit (10,000 km above the earth's surface). The system will be designed so that at least two HEOSs—which orbit the earth more slowly than LEOSs—will be in view of the user at all times.

9.5.4 Iridium

Iridium is Motorola's grand scheme for worldwide cellular telephone, paging, and data services. In particular, Iridium targets underdeveloped and remote areas of the globe. It was originally conceived as a fleet of 77 satellites and takes its name from the element with the same number of electrons orbiting its nucleus. The network has since been scaled back to 66 satellites, but the name Iridium has been retained because the element possessing 66 electrons is called dysprosium!

Subscribers will use mobile terminals, handheld units, or phones installed in solar-powered phone booths to communicate via the nearest satellite to their home gateway. The home gateway will track the user's location, monitor usage for billing purposes, and maintain user profiles (e.g., subscription rights). Iridium phones are expected to enter the market at approximately $3,000 each. Voice and data calls (up to 4,800 bps) will be billed at approximately $3 per minute. A seven-satellite pilot network is scheduled for launch by the end of 1994 (Figure 9.3).

User channels, each running 2,400 bps, will be derived from 31.5-kHz-wide TDMA channels and may be combined for higher speed services (for example, voice requires 4,800 bps). Mobile links will operate in the 1,610- to 1,626.5-MHz band; earth station links in the 20- to 30-GHz range. Each satellite will obtain its 590W of peak power from its batteries, which will be recharged as it travels over the oceans, where little traffic demand is anticipated. A phased-array antenna will project 48 cells, or interlaced spot beams, onto the earth's surface. A 12-cell reuse pattern will

Figure 9.3 Prototype dual-mode phone will operate over both Iridium and terrestrial cellular networks.

be employed, with each cell covering an area 200 to 300 miles in diameter. Active cells will total 2,150—about 68% of Iridium's worldwide capacity—and the system will deliver 4,720 user channels over the continental United States (CONUS, as it is known in satellite parlance).

Iridium will function like a terrestrial cellular telephone network—with one important difference. The motion of cells, rather than users, will cause cell handoffs. Unlike terrestrial cellular, Iridium's handoffs will be predictable and certain, and will occur about once per minute.

Iridium's satellites will be launched to polar orbits at an altitude of 780 km. The maximum distance between the user and nearest satellite will be about 2,300 km. The 66 satellites will be interconnected via microwave crosslinks, and each satellite will communicate with two satellites in front and two in back. An advantage of intersatellite crosslinks is that they reduce the number of terrestrial gateways required (15 to 20 Iridium gateways are planned) (Figure 9.4).

Iridium's initial launch is scheduled for 1996, and full deployment is expected by 1998. The satellites are designed to last 5 to 8 years, spares will be parked in space (using onboard propulsion systems to move into specific orbit locations), and emergency replacement satellites are promised to be launched within 36 hours. Khrunichev Enterprise of Russia will put 21 of Iridium's 66 satellites into orbit.

Motorola's primary goal is to create a global market for Iridium terminals. The firm plans to develop a handheld transceiver that will operate for up to 24 hours on a single battery charge (assuming 1 hour of talk time and 23 hours in standby mode). A dual-mode transceiver will always scan for terrestrial cellular channels first, switching to satellite frequencies if none are found. Motorola's Paging Group is also developing Iridium-compatible products.

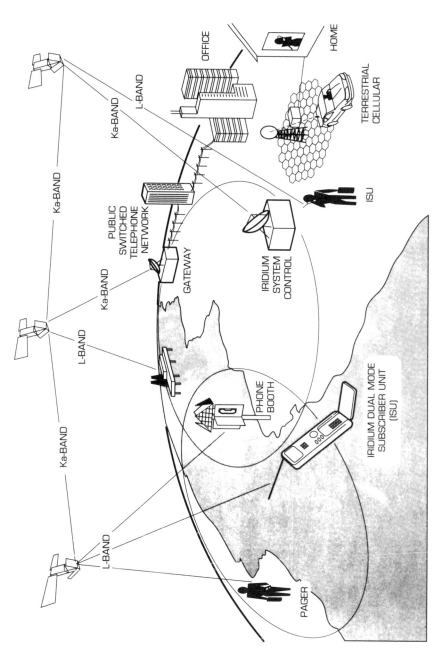

Figure 9.4 Iridium network will use satellite crosslinks to minimize the number of terrestrial gateways.

Iridium is a praiseworthy endeavor, but it faces a number of economic and political challenges. It is estimated that the Iridium network will cost in excess of $3 billion to construct. Iridium is making deals to head off concerns about competition for local PTTs and cellular carriers. Inmarsat members see Iridium as a U.S. manufacturer's plot to steal their market. In an attempt to allay such fears, however, Iridium has been courting international investors with the goal of significantly reducing Motorola's share in the venture. Iridium emphasizes that it is complementary to, and compatible with, terrestrial cellular networks. But with different terrestrial cellular standards in Europe, Japan, and the United States, there will always be concern that Iridium is really a "one-phone" solution for international business travelers.

9.5.5 Orbcomm

A Dulles, Virginia-based division of space launch vehicle manufacturer Orbital Sciences Corporation, Orbcomm proposes to construct a LEOS network consisting of 26 satellites in circular orbits approximately 600 miles above the earth, at a cost in excess of $250 million. From an initial approximate orbit, onboard propulsion will deliver and maintain each 87-lb MicroStar satellite to its precise orbit (Figures 9.5 and 9.6).

Orbcomm claims its LEOS design reduces the required transmit power by a factor of 150 to 1,000 compared to geostationary satellites. At Orbcomm's low altitude, VHF and UHF (as opposed to microwave) frequencies may be used, enabling smaller, more economical mobile terminals. The mobile downlink will run at 4,800 bps, and the mobile uplink at 2,400 bps. Orbcomm claims a basic VitalNet message communicator will be about the size of a pocket calculator, will use a whip antenna, and will sell for between $100 and $350. Annual subscription fees will cost $5 to $45 per year.

Unlike voice-based MSS, Orbcomm believes it can provide 100% coverage on urban streets. The canyon effect is overcome by the fact that its satellites travel across the sky and Orbcomm is a store-and-forward messaging service. The firm does not, however, promise inbuilding service.

Orbcomm will use doppler frequency shift for position determination. (That is, it will monitor changes in frequency as the satellite moves toward, and then away from, the subscriber device.) The drawbacks of this approach are that multiple readings are required for each location fix, and accuracy is only ±375m. Orbcomm satellites will house GPS receivers for computing their own locations.

Twenty 15-kHz-wide channels will provide 2,400 bps subscriber uplinks in the 148- to 150.05-MHz band. Each terminal will be randomly assigned a single channel at the time of manufacture for access on a contention basis. Handheld terminals will transmit at 2W, while vehicular mobiles will run at 5 Watts. Satellites will broadcast to mobile users at 4,800 bps over eight 27-kHz-wide channels in the 137- to 138-MHz

Figure 9.5 Prototype Orbcomm subscriber transceiver being developed by Panasonic.

band. Time and position information will be broadcast over a 50-kHz-wide UHF channel near 400 MHz. Orbcomm plans to construct nine regional gateways in the United States to communicate at 56 kbps over 100-kHz-wide channels.

Emergency search and rescue is one of Orbcomm's major target markets. Life-saving capabilities will be accessible nearly 24 hours per day, seven days per week, from any point on the globe. (There will be occasional brief service interruptions in some locations.) Hikers and mountain climbers will carry small terminals for signaling their positions. The position determination process takes 7 min to complete. Orbcomm's 24-hour-per-day, 365-day-per-year customer service center will notify the appropriate authorities. Orbcomm even offers a service that can detect a user who has fallen unconscious due to an accident. The user carries the unit in his or her backpack with antenna extended; failure by the user to press the "reset" button by an appointed time is interpreted as an indication something is wrong.

Automobiles will be equipped with dashboard terminals (using the vehicle's existing AM/FM antenna) to request emergency roadside service. Automotive terminals may be triggered by deployment of the vehicle's air bag, signaling both an

Figure 9.6 Orbcomm's gateway earth station enjoys a clear view of the sky from the high plateau near St. Johns, Arizona.

accident and the car's location, or remotely activated to track the vehicle if it is stolen.

Other interesting uses include worldwide paging and cellular telephone message taking. But to succeed, Orbcomm will need to attract hundreds of thousands of subscribers. Orbcomm's 2-min average transaction time will limit the number of suitable applications, but that will probably not be a major obstacle for international travelers who normally communicate across multiple time zones.

9.5.6 Qualcomm

The most successful land mobile MSS operator is Qualcomm, whose OmniTRACS service dominates the long-haul truck fleet management market. Qualcomm was able to achieve early success by adapting Ku-band satellites (originally designed for fixed VSAT applications) to mobile requirements. In a nutshell, Qualcomm brought a sophisticated new service to the market without the massive capital outlay and time delay required by proprietary satellite systems.

The use of satellites designed for fixed-site communications was made possible by Qualcomm's development of an electronically steered, mechanically driven mobile antenna, combined with advanced signal processing techniques. Except during and immediately after a sharp turn, the domelike antenna remains locked on its target.

OmniTRACS uses Ku-band transponders on GTE Spacenet's GSTAR 1 satellite. All user traffic passes through Qualcomm's 24-hour-per-day manned network control center. The satellite-to-mobile link runs at 10,000 to 15,000 bps using TDM, while the mobile-to-satellite link runs at 55 to 165 bps and uses CDMA. As the ranges imply, transmission speed is dynamically adjusted in response to received signal strength. The mobile downlink operates between 11.7 and 13.2 GHz, while the uplink is located between 15.0 and 15.5 GHz.

Qualcomm faces an obvious threat from new satellites designed specifically for MSS, which will be able to offer a wider range of services and better performance. Qualcomm has teamed up with Loral to develop a big LEOS called Globalstar. Globalstar is CDMA-based, consists of 48 planned satellites (each acting as a simple transponder), and is expected to cost $1.8 billion. Fax, voice, and e-mail will be available at $0.30 per minute. The Globalstar team predicts subscriber terminals will be priced at $750 each. (At that price, we may see dual-mode MSS phones subsidized, much the same as cell phones are.)

9.6 GLOBAL PERSONAL COMMUNICATIONS

MSS is a high-risk business. Inmarsat has succeeded because it has had no competition and, consequently, has been able to levy exorbitant usage charges. For the near term, MSS will continue to serve compelling applications like truck fleet management. It is not clear whether frequent international business travelers will use a mobile phone rather than a fixed phone to reach their home offices. (One clear advantage of MSS in this application is that it will make it easier for the home office to reach the global traveler.) And while MSS will bring modern telecommunications to the remotest parts of the globe, it is not clear that many people in such places will be able to afford the service.

LEOSs will become the platform for an array of global personal communications services. Until now, the lack of ubiquitous coverage has limited the acceptance of mobile communication services. LEOS solves that problem. In conjunction with urban terrestrial networks optimized for inbuilding coverage, mobile satellite networks will be able to offer true anytime, anywhere communications—the Holy Grail of personal communications.

SELECT BIBLIOGRAPHY

[1]   Adams, J. D., "Iridium Global Cellular Service," presentation at Wireless User '94 Conference, Orlando, Florida, March 23, 1994.

[2]   Clarke, A. C., *How the World Was One*, Bantam Books, 1992.

[3]   Brodsky, I., "Mobile Satellite Services," *The Wireless Industry Prospectus*, Datacomm Research Co., 1992.

[4]   Gilder, G., "Telecosm Ethersphere," *Forbes ASAP*, October 10, 1994.

[5]   Martin, J., *Telecommunications and the Computer*, Englewood Cliffs, NJ: Prentice-Hall, 1990.

[6]   Parker, A., "The First Global PCS Network," presentation at Wireless User '94 Conference, Orlando, Florida, March 23, 1994.

[7]   *Telecommunications Reports Wireless News*, various issues, BRP Publications, Inc., 1993, 1994.

[8]   Walker, J., ed., *Mobile Information Systems*, Norwood, MA: Artech House, 1990.

CHAPTER 10
▼▼▼

WHERE AM I?

10.1 RADIO LOCATING: THE NEXT UTILITY?

Rather than enabling communications anytime, anywhere, the main purpose of personal communications will be furnishing location-specific services. A new utility—the mobile information network—will supply data about our surroundings: Where am I? How do I get to my next appointment? Where should I take my client for lunch? Where is the nearest store that carries dress shirts with 38-inch sleeves?

Determining the mobile subscriber's precise whereabouts is central to such services. The ability to locate and track people, vehicles, and other objects has improved dramatically over the last few decades. Applications as far-flung as recreational boating, warfare, freight hauling, and personal safety are being served. Radio-locating technology has become progressively cheaper (under $500), ubiquitous (worldwide), and accurate (as good as ±10m).

Radio locating is a natural complement to mobile data communications. For organizations, tracking vehicles and personnel is essential to improving field productivity. Radio locating is a basic component of computer-aided dispatching. For individuals, radio locating combined with two-way mobile data delivers exciting new capabilities such as personal navigation (with access to real-time traffic data), emergency roadside service, and stolen vehicle recovery.

One technology in particular has captured the imaginations of both users and developers: the satellite-based GPS. Accessible by individuals equipped with low-cost handheld receivers, GPS provides amazing accuracy. So accurate, in fact, that the U.S. military purposely degrades the performance available to civilian users.

GPS achieved widespread notoriety by helping the Allies coordinate the movement of ground forces in the featureless desert during the Persian Gulf War. (Some believe, however, it also helped Iraq aim its SCUD missiles.) But GPS promises much more; combined with PDAs, notebook computers, and dash-mounted terminals, GPS receivers can provide turn-by-turn driving directions, location-specific advertisements, or quietly alert and direct law enforcement officials to a crime in progress. This is just the start; the introduction of PCMCIA GPS receivers will inspire software developers to create myriad new applications.

10.2 A BRIEF HISTORY OF RADIO LOCATING

From the beginning of recorded civilization, people have sought adventure, goods, and power through travel. Few have been bold enough to embark on long journeys with no means of finding their way home. Even fewer would eschew the opportunity to retrace their steps or show others the way to exotic destinations.

The Chinese used magnetic compasses as early as 1100 AD. Most of history, however, has been dominated by celestial navigation. Its chief tools have been the sextant, used to measure the angle between a celestial body and the horizon, and the chronometer. With the aid of these devices, one can compare the sun's actual position in the sky at a specific time of day to its expected position at a reference location (usually an astronomical observatory). Because of their distance, the stars appear fixed in space and provide an even more accurate guide. The configuration of Jupiter's moons served as an early chronometer, later replaced by clocks designed to withstand the jostling caused by the ocean's waves.

Each of these methods requires painstaking and time-consuming measurements—not to mention good visibility. While they can tell you which way to go, they offer no quick means of determining whether you are on course.

The wireless changed everything. Radio beacons were used to broadcast time signals for correcting shipboard chronometers. Beacons also made it possible to determine direction and later location, regardless of weather. In the early 1900s, beacons were situated at known locations, providing directions to maritime users using steerable antennas. By determining the direction of two or more known beacons, users could plot their position. But these systems were not without their drawbacks. Due to reliance on angular measurements, accuracy decreased with distance from the beacons.

Radar was the next major development. Invented by the Allies during World War II, radar provided accurate measurements of direction and distance, but early systems did not work much beyond the horizon. Radar was particularly useful in assisting aircraft landings during heavy fog, precipitation, or at night.

Dead reckoning was developed as a backup navigation system for use beyond radio range or under heavy jamming. Inertial navigation systems use accelerometers to measure changes in velocity and gyroscopes to measure changes in direction; after

World War II, they were sufficiently refined to serve as guidance systems on ballistic missiles. But all dead-reckoning systems suffer from the accumulation of error over time. For example, aircraft inertial systems drift 1 to 2 km per hour.

Beacons have been installed on aircraft and ships to aid search-and-rescue operations (mandated in 1972 after the disappearance of a small plane carrying two U.S. Congressmen over Alaska) and in automobiles to assist in the recovery of stolen vehicles. The alarming growth of U.S. auto thefts has prompted several firms to offer stolen vehicle recovery services using concealed beacons that may be remotely activated.

Wireless systems based on accurate timing signals have been developed for longer range navigation. There are two major types: those that measure the difference in time of arrival of signals from multiple sources, and those that measure the absolute propagation time from each source. The first approach is used by the LORAN C system of beacons and has served as the maritime navigation workhorse. The second technique is employed by the GPS, which now threatens to render all competing systems obsolete.

10.3 RADIO-LOCATING TECHNOLOGIES

Most short-range radio-locating systems employ *triangulation*. This technique may be used by a single receiver direction-finding multiple beacons to determine its own location, or by multiple receivers direction-finding a single beacon to locate the beacon (as used in search-and-rescue applications). Triangulation works by completing a triangle: once the length of one side of a triangle and the angles formed between it and the other two sides is determined, it is possible to plot the point at which the other two sides intersect (Figure 10.1).

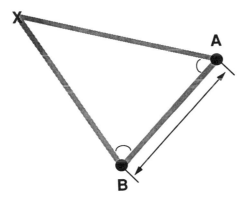

Figure 10.1 Two receivers equipped with direction-finding antennas at known coordinates (*A* and *B*) can plot the location of a beacon (*X*) using the triangulation method.

In contrast, most long-range radio-locating systems exploit the fact that the time required by a radio signal to travel between two points is proportional to the distance between those two points. For example, we can determine our relative position between two points by comparing the time of arrival of signals from the two points. Then we can plot a hyperbola (or straight line if the signals arrive at the exact same time) along which we must be located. If we repeat the process using a second pair of transmitters, we can plot a second hyperbola. At the intersection of the two curves is our location. This process is known as *trilateration* (Figure 10.2).

Alternatively, by measuring the (absolute) time required by a signal to reach us from a transmitter whose position is known, we can narrow down our location to somewhere on the surface of a sphere (velocity × time = distance). If we then measure the time it takes a signal to reach us from a second transmitter, we can identify a second sphere; our location is somewhere on the circle formed at the intersection of the two spheres. A third sphere allows us to narrow down our search to two points, one of which can usually be discarded as absurd (i.e., it might place us deep within the earth's crust, in the wrong hemisphere, or in outer space). If we are on the earth's surface, a geodetic database may be substituted for the third sphere.

Fortunately, microprocessors and semiconductor memories have enabled the development of devices that can automatically calculate one's position by manipulating equations representing these geometric objects. These devices have become increasingly smaller and less expensive. For example, there are now complete GPS receivers housed on credit card–sized PCMCIA PC cards (with external antennas).

Today, several radio-based navigation systems are in use. In the future, GPS may

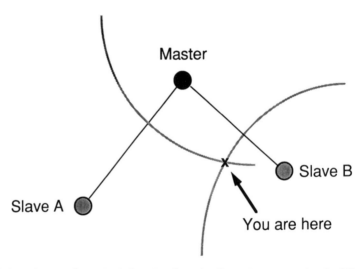

Figure 10.2 A receiver can determine its location through trilateration—measuring the difference in time of arrival of signals between two slaves and a master beacon.

become a standard for specific applications. There will continue to be a need, however, for remotely activated search-and-rescue beacons on ships and aircraft and concealed beacons worn by individuals. There will also be niche applications for networks that enable organizations to both track and communicate with vehicles in the field (see Chapter 11).

10.3.1 Beacons

With over half a million users, fixed beacons are both the oldest and most prevalent radio-based navigation aid. Maritime beacons operate at low frequencies, work at ranges of up to 300 miles, and offer accuracy of ±3 deg. Thanks to their proximity to the consumer AM radio band, beacons offer a low-cost solution.

Mobile beacons are used in search-and-rescue operations. Collaboration between the United States, former Soviet Union, Canada, and France led to the creation of COSPAS/SARSAT, which takes its name from the Russian COSPAS (Space System for Search of Vessels in Distress) and U.S. SARSAT (Search and Rescue Satellite-Aided Tracking System). COSPAS/SARSAT employs a combination of satellites and ground stations to locate downed aircraft, capsized boats, and other vessels in need of emergency assistance.

COSPAS/SARSAT consists of three Russian navigation satellites and three U.S. National Oceanic and Atmospheric Administration (NOAA) weather satellites. Emergency locator transmitters on aircraft are triggered by the impact of a crash, and floating emergency position-indicating radio beacons aboard ships are actuated by immersion in water. Distress signals are retransmitted to one of 20 ground-based local user terminals; the digitally encoded distress signal identifies the vessel's home base and owner. The user terminal relays location data to a mission control center from which search-and-rescue operations are coordinated. Since becoming operational in 1982, COSPAS/SARSAT has helped save over 2,000 lives.

New *personal beacons* have also appeared. The FBI has outfitted undercover agents with concealed beacons for years. Stolen-vehicle recovery services (SVRS) use beacons concealed in luxury cars. Airtouch Teletrac has proposed beacons disguised as wristwatches or other items that may be worn by children or elderly people afflicted with Alzheimer's disease. (Can the *implantable* personal beacon be far behind?)

10.3.2 LORAN

Measuring the difference in time of arrival of two signals was first proposed as a radio-locating technique in 1940 by National Defense Research Council (NDRC) section chief Alfred Loomis. The idea was attractive to the military because no transmitters were required on the vessel; only those on board would know its precise location.

The concept was first embodied in a system developed at MIT's Radiation

Laboratory to aid convoys crossing the North Atlantic during World War II. Dubbed long range navigation (LORAN), the first network operated on 1.95 MHz, used ground wave propagation out to 1,200 km during the day and sky wave propagation out to 2,000 km at night, and was accurate to within ±1.5 km under all weather conditions.

At the conclusion of World War II, the U.S. Coast Guard took over the network, renaming it LORAN A. It provided position fixes by measuring time differences between either three synchronized stations or two pairs of synchronized stations. Several variants of the LORAN system also appeared. Omega was designed to provide worldwide coverage via fewer transmitters. It uses continuous-wave (CW) signals rather than pulses, measures differences in phase angles as opposed to time, and operates in the 10- to 14-kHz band. Although Omega's accuracy is only ±2 to 4 miles, it covers the entire globe with base stations located in North Dakota, Argentina, Australia, Hawaii, Japan, Liberia, Norway, and Reunion Island.

A drawback of the early Omega and LORAN systems was that they were two-dimensional only. LORAN C was developed to provide ±100m accuracy for tactical aircraft and was used during the Vietnam War. Today, LORAN C operates at 100 kHz, serving the continental United States, the North Atlantic, North Pacific, and Mediterranean through over 50 beacons. Although deployed in groups of three or five, stations within a group are typically located hundreds of miles apart; accuracy varies with distance from the beacons. The network has continued to expand, now covering the entire land mass of the continental United States. LORAN C has approximately 400,000 maritime users, but faces competition outside the United States from the Decca system. (Decca uses CW technology, operates in the same frequency band as LORAN C, and leases receivers to users.)

10.3.3 Transit

The first satellite-based radio navigation system was jointly developed by the U.S. Navy and Johns Hopkins University's Applied Physics Laboratory (APL). Called Transit, it is a dual-frequency (150 and 400 MHz) system based on doppler frequency shift measurements (the mobile receiver determines its location by observing the frequency shift from satellites in known orbits as they pass overhead). Although Transit's accuracy is as good as ±30m, the satellites are only in view an average of once per hour from any given location.

Transit's origin is interesting. In 1957, APL scientists demonstrated that they could determine the ephemeris (locations over time) of the Soviet Union's Sputnik I by measuring the doppler shift of its CW transmitter. It soon became apparent that the inverse was also true: if one knows the satellite's ephemeris and measures doppler shift as it passes overhead, it is possible to calculate one's position on the ground.

Transit was also the first system to introduce ionospheric propagation velocity corrections. This was done by observing the difference in time of arrival of the two signals (150 and 400 MHz), from which it is possible to calculate the delay solely attributable to the ionosphere.

The first Transit satellite was launched in 1959, but the system did not become fully operational until 1968. By 1990, Transit consisted of seven operational satellites (and six spares "parked" in space) and had close to 90,000 users. When a Transit satellite is in view, fixes may be obtained every 30 sec. Satellites pass overhead intermittently: every 30 min at 30° north or south latitude to a maximum of every 110 min at the equator; a single satellite may be in view for up to 20 min. Accuracy is degraded for users in motion. Due to its slow, intermittent, and two-dimensional-only service, Transit is expected to be gradually phased out in favor of GPS and the Russian GLONASS system.

10.3.4 Navstar/GPS

The idea behind the GPS emerged in a 1960s study examining future military aircraft and ballistic missile navigation systems. The U.S. Air Force's Space Division decided to back the development of such a system in 1963. Its top priorities were the ability to obtain continuous position fixes anywhere in the world, three-dimensional accuracy to support bombing missions, a system that required no radio transmissions by mobile users (potentially giving away *their* position), and automatic operation.

In essence, GPS replaces the stars with satellites. Using radio instead of light waves, GPS works under all weather conditions. Just as stars serve as known points in the sky that may be seen, GPS provides the precise locations of points (satellites) in the sky that may be heard.

A team consisting of Aerospace Corporation, IBM Federal Systems, The Global Positioning Team, Rockwell International, the U.S. Air Force, and the U.S. Naval Research Laboratory designed the final GPS system. The project was renamed Navstar (Navigation Satellite Timing and Ranging) in 1974, but the abbreviation GPS refused to die. Construction commenced in 1989 and will be completed by 1996. Initially granted $150 million in funding, GPS has mushroomed into a multibillion dollar project. Although not fully operational at the time, the power of GPS became famous during the Persian Gulf War.

What began as a military system has increasingly found civilian uses. Today, GPS is used for commercial maritime and aircraft navigation, surveying and geological studies, the distribution of precise worldwide time to synchronize digital communications, and recreational boating. Since there are no costs associated with using GPS, affordability is strictly a function of receiver cost. (At one time, the U.S. Congress attempted to impose user fees, but the move failed due to fears of inhibiting applications involving public safety.)

GPS operation differs from Transit, LORAN C, and related systems. While LORAN C measures differences in signals' time of arrival and Transit measures frequency shifts, GPS measures the absolute time it takes signals to reach receivers on earth from satellites in known orbits. While GPS is more complex, it offers greater accuracy and speed. In the end, complexity always yields to silicon integration; small GPS receivers can now be purchased for well under $500.

Because LORAN C receivers make only relative time measurements, they can use inexpensive crystal oscillators. To measure the precise time required for a signal to travel from a satellite to a mobile user, however, requires extremely accurate and perfectly synchronized clocks at both ends. At first, this requirement appears overwhelming. While it is not unreasonable to equip multimillion dollar satellites with $100,000 atomic clocks, they are out of the question for end user devices.

Hideyoshi Nakamura of Aerospace Corporation invented a mathematical solution to the problem. He realized there are four unknowns involved: the user's distance from each of three satellites, plus the receiver clock error. By obtaining a fourth satellite reading, the four unknowns can be solved using four simultaneous equations.

A two-dimensional example helps demonstrate how this mathematical device works (Figure 10.3). If we are certain there is no clock error, we can pinpoint the location of a user in a two-dimensional universe using just two satellites. The distance from each satellite defines a circle. The intersection of two circles generally yields two points. (We will assume that even in a two-dimensional world, one of the two points is immediately recognized as absurd and may be discarded.) We now have one point which we believe is the true location.

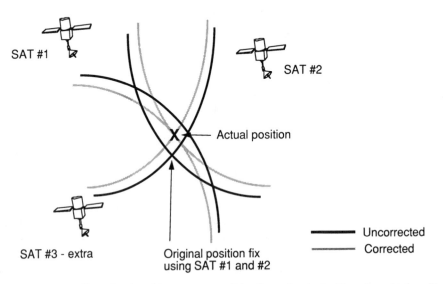

Figure 10.3 In a two-dimensional world, an accurate position fix can be obtained by using a third satellite measurement to correct for GPS receiver clock error; the same can be done in our three-dimensional world by using a fourth satellite.

Now suppose there was clock error. Although the error affected both of our measurements, they still yielded two intersecting circles. We had no way of determining whether, and to what extent, our position fix was inaccurate. Still operating in a two-dimensional world, imagine we obtain an extra (third) satellite reading. Like the first two, the new measurement defines a circle, also thrown off by clock error.

Had there been no clock error, the third satellite measurement would have proven redundant, simply reconfirming the original position fix. Instead, the third circle intersects the first two circles at points that do not coincide with our original position fix. The good news is that the receiver clock error is the same for all three measurements. If we incrementally add positive and negative error estimates to our original measurements—beginning with small values—we will find error values at which the three circles will begin to converge. We can continue this process until the three circles intersect at just one point. We now have the correct location fix and an accurate clock error measurement. An analogous procedure, involving a fourth satellite reading, may be performed in our three-dimensional world.

Users may also lock into universal coordinated time (UTC) through GPS. (UTC is an international time standard that harmonizes the second, as defined by atomic clocks, with the calendar, as defined by the earth's motion around the sun.) Once a GPS receiver calculates its precise location, as described above, it automatically knows its exact distance from the satellites currently in view. To determine the exact time, the receiver simply subtracts the calculated propagation delay from the received (apparent) time value. These time measurements may be used to synchronize digital communication networks throughout the world.

In order for GPS receivers to receive signals from multiple satellites simultaneously, each satellite's antenna must spread its approximately 20W of transmit power over an entire hemisphere. As a result, the GPS receiver sees a signal 30 dB below the ambient noise level! When this 10-MHz-wide spread spectrum signal is correlated back down to 100 Hz, however, there is a 50-dB improvement—with no increase in ambient noise. In addition, the use of CDMA—first proposed for GPS in 1965—enables all of the satellites to transmit on the same frequency. (A multichannel GPS receiver, therefore, is not listening on different frequencies, but to different logical channels.) Due to growth in civilian use, GPS is one of the first commercial successes for CDMA.

GPS transmits two simultaneous (90-deg phase-shifted) spreading codes: the coarse acquisition (C/A) and precision (P) codes. The C/A code is 1,023 chips in length and is repeated 1,000 times per second. The P code consists of 40 seven-day segments containing 6.5 trillion chips apiece; each satellite is assigned a different segment. As its name implies, the P code provides more accurate positioning, but the C/A code is quite accurate—even when used alone.

The final GPS design—a compromise between the different branches of the U.S. military—consisted of 24 satellites in 12-hour orbits at an altitude of 10,900 miles. At least four satellites are visible to any user at any time. Critics pointed out that user terminals could be jammed, the satellites could be attacked in space, and—perhaps most disturbing—enemies could also use GPS. Addressing the latter concern, the Air Force decided to develop a means of introducing deliberate time and satellite position errors, called selective availability (SA), which would reduce unauthorized (civilian or foreign) users to ±100m accuracy. This level of accuracy was deemed sufficient for navigation, but insufficient for delivering weapons to their targets.

In order to operate properly, GPS receivers must possess the daily ephemeris and time correction data, which can be either preloaded or acquired from any GPS satellite. (This information is only available looking out about six months.) When evaluating the time it takes a GPS receiver to calculate a position fix, it is necessary to distinguish between the time to first fix (perhaps 2 min) and subsequent fixes (which can be computed about twice each second).

All GPS satellites operate on the same mobile, earth station, and intersatellite frequencies. The mobile frequencies are known as L1 (1,575.42 MHz) and L2 (1,227.6 MHz). These frequencies are exact multiples of the atomic clock's 10.23-MHz output. Cesium atomic clocks stable to one part in 10^{13} are used, with clock drift corrections uploaded as necessary from master ground stations.

Errors in satellite position, time, and ionospheric propagation data are common to all users in the same geographic area. If one can correct these errors for one user, therefore, it is possible to correct them for all users viewing the same group of satellites at the same time. In fact, this is what many users do, and it can even be used to defeat SA. By installing a master receiver at a carefully surveyed location, the necessary corrections may be derived and relayed to other users. This technique is known as differential GPS (DGPS).

DGPS can improve accuracy to within a few centimeters. It is used to enhance navigation near airports and harbors. DGPS is also being used to measure shifts in the earth's crust. There are several ways to deliver DGPS corrections—everything from FM subcarrier to private mobile radio.

GPS is quickly becoming an international standard for maritime and aeronautical navigation. It is also working its way into land mobile applications like long-haul truck fleet and commuter transportation system management. There are reportedly over 150 manufacturers of GPS receivers. Some believe GPS will eventually be spun out of government hands, but it is not clear how the service would make money, especially in civilian applications.

Will GPS be adapted to notebook computers, PDAs, and personal communicators? It already has—Bel-Air Technologies (San Jose, California) sells the 2.4-lb Vista-GPS portable PC with an integrated GPS receiver. But further price reductions will be necessary to make PDAs with plug-in GPS receivers a consumer item. And consumers will expect a wide range of software and services to become available—everything from instant driving directions to detailed information on local businesses.

10.4 GEOGRAPHIC INFORMATION SYSTEMS

Geographic information systems (GIS) gather, store, organize, and display geographic data. This may include the locations of customers, roadways, and city infrastructure; consumer demographics; and real-time data regarding the coordinates and status of field personnel and vehicles. In essence, the GIS is a platform for scheduling, routing, and other tasks involved in managing an organization's field operations.

The electronic map is the most common user interface to the GIS. Electronic maps are computer versions of printed maps—only much more powerful. They not only provide a graphical representation of transportation infrastructure, they allow the user to manipulate and display data in customized ways. Most electronic maps are used to manage fixed resources, but a growing number are equipped with real-time interfaces for managing mobile resources. For example, a real-time interface enables one to display the current location and track the movement of a repair van.

The first geographic information systems were introduced in the early 1970s. These systems were designed to run on mainframe computers and were, therefore, too expensive for widespread adoption. Most of the early users were government agencies and large corporations. The development of powerful desktop computers—coupled with progress in relational databases—has expanded GIS's usefulness to medium- and small-sized users. A number of GIS products are now available that run under Microsoft's Windows or Apple's Macintosh operating system.

10.5 GIS TECHNOLOGIES

Geographic information systems combine graphics, computer-aided design, and database management software. To be used effectively, most require extensive experience in database management. Fortunately, many have been customized for computer-aided dispatching, making life simpler for end users primarily interested in routing and scheduling.

The coverage of an electronic map may range from a single city block to the entire world. They may contain items ranging from streets and pipelines to information on natural terrain and crop locations. Most electronic maps allow the user to zoom in or out of various viewing levels, and to select or deselect the types of items to be displayed. In truck fleet management applications, for example, one might zoom back and forth between regional and local views.

There are two major types of electronic maps. A digital electronic map is constructed solely from a database. Digital maps tend to lack texture—particularly the kind that is pleasing to the eye—but are easy to modify. Analog electronic maps are created by scanning printed maps and matching the image to computer-generated coordinates. Analog maps are usually richer in visual information, but may be more difficult to modify, since some of the information displayed is not contained in any database (Figure 10.4).

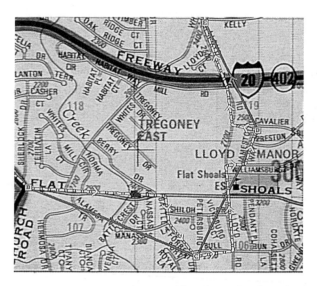

Figure 10.4 Electronic maps, such as this one from Roadshow International, can be created by overlaying digital coordinates on images scanned from richly detailed printed maps.

Analog maps possess a key advantage: one can quickly computerize virtually any area with an associated printed map. Digital maps, on the other hand, require painstaking data collection. It is no small effort to create and maintain geographical databases for all major U.S. cities. New construction tends to make the databases obsolete within months or even weeks.

10.6 RADIO-LOCATING APPLICATIONS

Since it was invented, the main applications of radio locating have been navigating ships at sea and aircraft. Newer applications include precision mapping, missile guidance systems, and aligning drilling machines digging tunnels from opposite directions (such as the new tunnel under the English Channel).

Radio locating can serve enterprise applications, particularly when it is combined with a two-way mobile data link. Long-haul freight carriers, for example, can identify and reroute the closest truck to a new pickup order. The truck dispatch center can project a map of the country on a wall, replete with icons representing the present locations for each of its trucks. In the battle for the long-haul freight market, companies can enjoy all the benefits of a high-tech war room!

Cities such as Denver, Colorado, are installing GPS receivers connected to mobile radios to track commuter buses. In this application, the operations center receives continuous updates on the locations of buses and transmits estimated time of arrival information to electronic signs at bus stops. The operations center can also reroute buses around accidents or order off-schedule buses to switch to express mode. Once busses are equipped with two-way data communications, drivers can also be provided with a silent alarm for summoning the police.

Public safety agencies can use radio locating not only to guide emergency vehicles, but also to establish an emergency-response audit trail for later analysis. Other potential applications include transmitting position data from major accident sites, locating stranded motorists, and rescuing hikers who are lost or injured.

There are also consumer applications for radio locating. Just about anyone could benefit from turn-by-turn driving directions. However, providing good driving directions is harder than one might think. Directions given by people often include warnings about staying in the correct lane or watching for important landmarks.

Eventually, pocket-sized PDAs and personal communicators with electronic maps will use GPS to pinpoint the user's current location. Not only will users know where they are, but also where they have been. This could come in handy in getting back to a parked car or retracing steps to find a missing piece of jewelry. But a mass consumer market will require ultracheap GPS receivers—perhaps under $100.

10.7 PROFILES OF SELECTED VENDORS

10.7.1 Trimble Navigation

Trimble Navigation is a leading manufacturer of GPS receivers. According to the company, "Knowing where you are is so basic to life, GPS could become the next basic utility." This attitude differentiates Trimble from many of its competitors who see GPS as a specialized service, primarily for government and commercial navigation, vehicle tracking, and surveying.

Nevertheless, the bulk of Trimble's products target military, marine, aviation, surveying, and automatic vehicle locating (AVL) applications. For example, the Pathfinder is a portable 3W navigation and position logging system (including GPS receiver, data logger, and battery pack) that weighs under 10 lb. The GPS Navigator is a combination GPS receiver and LORAN C receiver for aviation applications.

Several of Trimble's GPS receivers feature communication ports for external display and analysis, or interfacing with mobile data networks for centralized locating and tracking of field workers and their vehicles.

The TRIMPACK (ruggedized for military requirements) and TransPak GPS are handheld three-channel GPS receivers with backlit displays. The TransPak measures $6.5 \times 7 \times 2$ inches and weighs 4.2 lb with its NiCad battery installed. The TransPak can determine three-dimensional position and altitude as well as velocity. Once it calculates the first fix, the unit can calculate an update every second. The TransPak features user-selectable land, air, and sea modes. The Ensign GPS handheld for boating was released in 1992 and includes a "man overboard" feature for marking the site of an accident. See Figure 10.5.

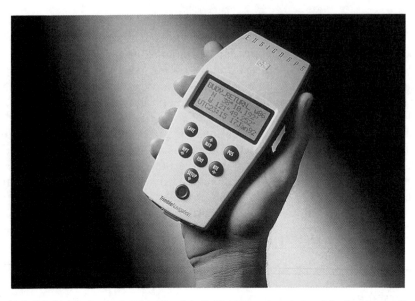

Figure 10.5 Trimble Navigation's Ensign is a handheld GPS receiver designed for boating applications.

Trimble announced what it calls the "World's first PCMCIA GPS product" in February 1993. The firm envisions applications like consumer guides, personal navigation, real estate appraisals and sales, market location analysis, insurance claims adjustment, public safety, and emergency disaster relief for GPS-equipped mobile computers.

10.7.2 Navigation Technologies

Navigation Technologies (Sunnyvale, California), also known as NavTech, is a joint venture between U.S., Japanese, and European automotive electronics firms. The company is constructing a navigable database of all major cities and intercity road networks throughout North America. Applications range from market research to fleet management.

The database includes street names, block-by-block addresses, speed limits, traffic signals, turn restrictions, and points of interest (e.g., hotels, restaurants, parks, and schools). NavTech's database is designed to accommodate ongoing maintenance (promising minimum accuracy of 97%) and customization. The firm has research personnel throughout North America and plans to input data on a geographical area representing 125,000,000 inhabitants by the end of 1995.

NavTech's investors and strategic partners include participants in the IVHS industry (see Chapter 11) and other companies developing geographic information and navigation products. These include the American Automobile Association, Mo-

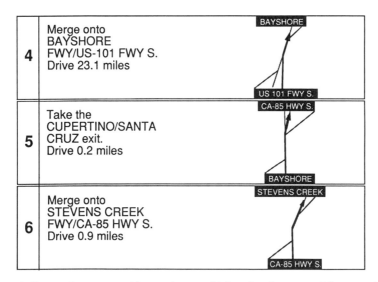

Figure 10.6 Online services can provide turn-by-turn driving directions to mobile users via two-way wireless links.

torola, Nippondenso Company, Philips Electronics N.V., SEI Technology Group, and Zexel Corporation.

NavTech bases its maps on existing data from aerial photographs, local (analog) maps, and other data sources. The firm believes that many of the original digital maps were assembled without sufficient regard for the requirements of ". . . sophisticated intelligent vehicle highway system applications." The database is composed of links, nodes, paths (sequence of related links), polygons (bounded areas), points of interest (POI), and value-added data (VAD; e.g., turn restrictions).

NavTech's database has been used in a traveler kiosk (for turn-by-turn driving directions between any two addresses within a city), for invehicle route guidance (see Figure 10.6), and in computerized production of printed street maps. NavTech has announced a strategic alliance with Sundisk to provide removable map data on PCMCIA-compatible solid-state mass storage cards. This will enable PDA and personal communicator users to consult city street maps and other traveler information. In conjunction with Motorola, NavTech will provide an online driving direction service (called DriverGuide) for the Envoy personal communicator. The user will simply input the origin and destination addresses. A map and step-by-step directions will then be transmitted back.

10.8 YOU ARE HERE

Trimble Navigation is right: radio locating will become an everyday tool for both business users and consumers. Discovering where you are and how to get somewhere else have become basic tasks in today's highly mobile society.

Despite the declining cost of semiconductor memory, it is a safe bet that our proficiency at collecting and generating information will grow at an even faster pace, and the personal communications revolution will facilitate our ability to access it. Although one can already purchase electronic maps on PCMCIA cards (e.g., *Fodor's '94 Travel Manager*), memory is no substitute for the functionality made possible by wireless data communications. Equipped with a GPS receiver and two-way wireless data, information appliances will not only display local maps and travel directions, but allow users to browse store shelves, check current traffic conditions, inspect today's lunch menu, and make reservations. A cottage industry will arise to create the content for new location-specific information services.

SELECT BIBLIOGRAPHY

[1]   Troy, E., "An Introduction to the GPS System," *Spread Spectrum Scene*, December 1993.
[2]   Getting, I., "The Global Positioning System," *IEEE Spectrum*, December 1993.
[3]   Hurn, J., "GPS: A Guide to the Next Utility," pamphlet published by Trimble Navigation, 1989.
[4]   Trudeau, M., "At Last, a GPS Experience," *GPS World*, July 1994.

CHAPTER 11
▼▼▼

SMART CARS, SMART ROADS

11.1 DRIVER INFORMATION SERVICES?

A new kind of road network is emerging—a network that will guide us to our destinations, alleviate traffic congestion, provide immediate access to emergency services, and even do some of the driving for us. Combining mobile communications and surface transportation technologies, the brave new world of intelligent vehicle highway systems (IVHS) may one day permit us to read the morning newspaper while commuting to work in our personal automobiles.

Nay-sayers contend that as traffic flows more smoothly, more people will take to the roads, which can only lead to new congestion. At its heart, this is an argument against industrial development; the same argument could have been made against constructing the very first roads. Those who believe in the merits of industrial development recognize that IVHS will improve the utilization of roads and, in doing so, will help ensure that modern surface transportation is accessible to everyone.

But a controversy is brewing over how the market for IVHS products and services will evolve. Some see a near-term opportunity to sell autonomous mobile navigation devices. Others believe a massive infrastructure is required to monitor and control traffic. Clearly, autonomous devices would be enhanced by access to infrastructure-provided services such as real-time traffic conditions. Which must come first—the infrastructure or autonomous devices—is a classic chicken-and-egg dilemma.

The advantage of starting out by constructing a powerful infrastructure (where most of the processing and data storage capabilities reside) is that it would enable drivers to use inexpensive terminals. But perhaps the main lesson of the personal computer revolution is that as technology develops it becomes more—not less—accessible to individuals. An installed base of autonomous devices would provide a ready market for future infrastructure-provided services. (Indeed, several major automobile manufacturers are already offering navigation systems in certain luxury car models.) These autonomous devices could be PDAs and personal communicators—perhaps used in conjunction with indash docking stations.

But who will pay for, construct, and operate the infrastructure? To many, the obvious choice is government. Others suggest that once standards are in place, the operation of such an infrastructure could be franchised, much like cable TV. Alternatively, automobiles equipped with autonomous devices and two-way mobile communications could form a *virtual infrastructure* (e.g., by serving as roaming traffic probes). But there are obvious advantages to integrated road sensors, beacons, and traffic signals, particularly in providing coordinated control. There is no reason why multiple systems—each focusing on different markets—could not coexist.

11.2 A BRIEF HISTORY OF INTELLIGENT VEHICLE HIGHWAY SYSTEMS

A 1986 Texas Transportation Institute study of 29 major cities estimated that the U.S. economy loses $24 billion annually to highway congestion. Of that total, over $1 billion is lost per year in each of the 12 largest cities. Costs to society include accidents, wasted fuel, and increased costs for monitoring and reducing auto emissions. Highway congestion is even linked to reduced worker productivity; time and energy people could apply on the job is wasted fighting traffic during the commute to and from work.

Japan is generally credited for launching IVHS in the mid-1980s with its $30 million radio/automobile communication system (RACS) project. RACS used one-way microwave beacons, spaced 2 to 10 km apart, to transmit traffic and location updates at 64 kbps along highways to vehicle-based dead-reckoning systems. Because it was designed to provide intermittent coverage, the beacons could operate on the same frequency (in the 2.5-GHz band) nationwide.

In 1988, a navigation experiment called AMTICS (advanced mobile traffic information and communication system) was conducted in Tokyo. (In Japan, city addresses are nonsequential and are instead based on area, block, estate, and parcel number; it is often difficult even for natives to find a specific address.) Today, a group of Japanese government agencies and private companies continue to pursue IVHS under a project known as vehicle information communication system (VICS), which combines RACS and AMTICS and calls for widespread deployment in the 2010s. But commercial automobile navigation systems are already enjoying success in Japan, and are available with cars from Honda, Toyota, Nissan, Mazda, Mitsubishi, and Suzuki.

In Europe, IVHS is known variously as road transport informatics (RTI) or advanced transport telematics (ATT). In 1986, 18 European car makers—along with the European electronics industry, telecommunications authorities, traffic engineering agencies, and research institutions—began work on a project primarily intended to improve the competitiveness of European-made automobiles. The PROMETHEUS project (Program for European Traffic with Highest Efficiency and Unprecedented Safety) consists of several different endeavors—including Pro-Car, Pro-Net, Pro-Road, Pro-Chip, Pro-Art (artificial intelligence), Pro-Com, and Pro-Gen—and has received a total of $700 million in funding.

A second major European project called DRIVE (Dedicated Road Infrastructure for Vehicle Safety) was launched in 1989 and focuses on the development of infrastructure. The primary objectives of the project's second phase (DRIVE II) are to improve safety, efficiency, and the environment. This includes research in areas such as public transport, fleet management, driver-assisted and cooperative driving, urban traffic management, and travel information.

Fearing that the United States was behind in IVHS research and would become dependent on foreign technology, a group of individuals representing U.S. government, education, and industry formed Mobility 2000 in 1989. Out of this organization, IVHS America was established the following year as an advisory committee to the U.S. Department of Transportation (DOT) to plan IVHS technologies and draft standards. Over 500 members strong, this organization is funded by federal, state, and local government, private industry, and academia. Its mission statement reads, "Coordinate and foster a public-private partnership to make the U.S. surface transportation system safer and more effective by accelerating the identification, development, integration, and deployment of advanced technology."

In 1991, the U.S. Congress established the Intermodal Surface Transportation Efficiency Act (ISTEA), authorizing $660 million to be spent on IVHS from 1992 to 1997, with the goal of putting into operation the first fully automated test track or roadway. Reflecting the growing sense that smart road technologies should not be limited to highways, IVHS America voted to change its name to ITS America (Intelligent Transportation Society of America) in September 1994.

Today, dozens of IVHS pilot projects are under way or under development in the United States, Japan, and Europe. There are over 40 such projects in the United States alone, including ADVANCE, a large-scale route guidance system in the northwest suburbs of Chicago; Pathfinder, a three stage project to evaluate invehicle navigation systems for use in the Los Angeles area; and TRAVTEK, a public/private partnership to provide motorist information and route guidance in Orlando, Florida.

The ADVANCE project is unique because it focuses on reducing congestion on suburban streets (rather than highways) over a 450-square-mile area. The major partners are the Federal Highway Administration (FHA), the Illinois Department of Transportation (IDOT), Motorola, Northwestern University (Evanston, Illinois), and the University of Illinois at Chicago. In Phase II of the project, 5,000 privately owned vehicles will be equipped with route guidance systems and will also serve as roaming traffic probes. The project team estimates that, with this number, it will be possible to monitor travel times every 10 min on 70% of major roads during peak weekday hours.

Routes will be selected based on driver preferences such as shortest time, shortest distance, or no expressways. The ADVANCE system will operate in two modes: General and Commuter. In the General mode, the vehicular computer will give turn-by-turn driving directions. In the Commuter mode, the system will present alternative routes based on current traffic conditions. Driver alerts will be presented in both modes.

ADVANCE will use a combination of dead reckoning, map matching (correcting cumulative tracking error by matching the vehicle's location and direction of travel with a known street), and GPS (which requires four satellites to be within view, but does not function in wooded areas, parking garages, or on some tree-lined streets). With a mobile navigation assistant (MNA), routes will be selected from a database containing historical information (such as travel time as a function of time of day). The vehicles will report current travel times to a central computer. It will also be possible for the driver to push an SOS button for emergency assistance. The driver will then be asked to select fire fighting, tow truck, police, or ambulance service.

In Europe, Philips is developing a system called Car Information and Navigation System (CARIN), which accesses map data from a compact disc and uses a speech synthesizer to present directions. SOCRATES (System of Cellular Radio for Traffic Efficiency and Safety) is another approach that uses an onboard database for autonomous route planning. SOCRATES is designed to work in conjunction with the GSM cellular network, using the downlink to broadcast traffic information and the uplink (employing a multiple-access technique) to gather information from vehicles on current trip times. In this way, just two GSM channels per cell could serve a large population of users.

Other navigation systems have been, or are being, developed that require a smaller investment in hardware per vehicle. Siemens' EURO-SCOUT (ALI-SCOUT in the United States) provides location information via infrared beacons installed at major intersections. Although this system (see below) requires more infrastructure, it facilitates collective route planning—a more proactive approach to traffic management.

11.3 FROM ROUTE GUIDANCE TO PLATOON DRIVING

IVHS is not a single service, but a family of passive (information) and active (controlled driving) services. Passive applications include navigation and onboard *electronic yellow pages* (i.e., information on local businesses, tourists spots, and hospitals). Briefly taking control of a vehicle when it comes perilously close to another—*collision avoidance* or *intelligent cruise control*—is likely to be the first active service deployed.

More advanced systems are on the drawing board that would assume near-total control of vehicles as they merge onto highways. Most notable is *platoon driving*, in which packs of vehicles would speed along spaced just a few feet apart. Some proponents advocate setting aside special lanes for platoons; they believe seeing cars speeding along in these lanes during rush hour will attract more participants. Thanks to computerized control, platoon lanes could also be narrower than conventional lanes, effectively widening roads without tearing them up.

The major types of IVHS services as originally defined by Mobility 2000 are:

- *Advanced public transportation systems* (APTS): Applications include automated fare collection and real-time transportation information. Several cities are using, or planning to use, GPS receivers and two-way mobile data links to track busses in order to provide waiting commuters with the estimated time of arrival (ETA) of the next bus, and to help plan or dynamically change bus routes.
- *Advanced traveler information systems* (ATIS): The purpose of ATIS is to provide driver information (navigation) and directions (suggested routing) in real time. These services may be provided either external to the vehicle (message signs) or via invehicle information systems. ATIS includes traffic information broadcasting, safety warnings, onboard navigation, electronic route guidance, and public transportation access information. Vehicles equipped with navigation systems plus a reverse data link could serve as traffic probes for a central computer monitoring travel times along key arteries.
- *Advanced traffic management systems* (ATMS): The purpose of ATMS is to improve traffic flow. ATMS includes urban traffic control systems, incident detection systems, highway and corridor control systems, and ramp metering systems. The hardware used to operate ATMS includes road sensors, traffic signals, ramp meters, electronic message signs, and communications devices. Systems may be designed to provide coordinated control over a wide area, or to simply control individual intersections.
- *Advanced vehicle control systems* (AVCS): The purpose of AVCS ranges from driver warning systems to automated braking, speed control, and steering systems. Some of these functions could be performed independent of the driver. While AVCS might take control of your vehicle when you merge onto a highway, it would (if all goes according to plan) relinquish it when you reach your preprogrammed exit. California's PATH experiment has tested platoons of four cars moving at 55 mph, in which three follower cars were slaved to a manually driven lead car using a vehicle-to-vehicle communications network.
- *Commercial vehicle operations* (CVO): Applications include electronic fee payment and hazardous cargo tracking. Radiolocation is used to track fleets of vehicles. RF identification is used for weigh-in-motion (WIM) and automatic vehicle classification (AVC) functions. Up until now, most IVHS pilot projects have focused on commuter needs, but there is also a demand for commercial

services designed to reduce the frequency of stops and eliminate the red tape associated with applications such as interstate trucking.

11.4 TALKING ROADS

A multitude of communications technologies can be used to enable smart cars and smart roads. They include both existing and new services, and both radio and infrared. Some services, such as cellular telephone, SMR, and AVM are particularly well positioned. Nationwide mobile data services—still struggling to find subscribers—see IVHS as one of many potential markets. Some have suggested data could ride piggyback over vehicle-based collision-avoidance radar signals. And infrared can be used both outdoors, for beacons, and within a vehicle to link handheld computers to in- or under-dash docking stations.

AVM is a strong candidate because it offers integrated vehicle tracking and two-way mobile data. There are two types of AVM services in the United States: those created for fleet management applications, and those that evolved out of the SVRS. Unlike radiolocation services that enable a mobile device to pinpoint its own location, AVM networks are designed to locate the mobile transmitter from a central site by comparing measurements from a series of base station receivers.

Nevertheless, the communications needs of IVHS have not yet been fully determined. IVHS researchers see communication requirements falling into four major categories: area broadcasting, local roadside beacons, two-way mobile radio, and local roadside transceivers. Area broadcasting could be used to download temporary or permanent changes to databases. These might include updates to electronic maps due to new construction (perhaps broadcast once or twice a day), or hazard warnings (broadcast in real time). Hazard warnings might specify the types of vehicles affected, the location of the hazard, and advice on what to do.

Cue Paging provides data broadcast services over commercial FM radio station subcarriers (Figure 11.1). A service called TrafficAlert transmits highway congestion information at 8,000 bps (and emergency voice alerts on a second channel) to users equipped with Sharp Wizard personal organizers attached to Cue's DataPlatform radio modem (Figure 11.2). Cue Paging received a $3.2 Million grant from the FHA to install a demonstration system between San Francisco and Reno. The driver keys in the highway number, the origin and destination, and direction of travel to initiate service. Cue's ComCard software contains a database of all major highway entrance and exit ramps within the service area. Transmissions indicate where (highway intersection or on/off ramp) traffic speed has slowed to below 35 mph. The advantage of TrafficAlert over conventional radio traffic reports is that drivers receive alerts only for their specific route (rather than listening to area-wide voice reports); a more detailed report can then be obtained by pressing the "Status" button.

Seiko Communications plans to build a global FM subcarrier network based on its 19,000-bps ACTTIVE™ technology. The firm has developed a miniaturized chip set that enables the integration of a pager in a standard-sized wristwatch (the

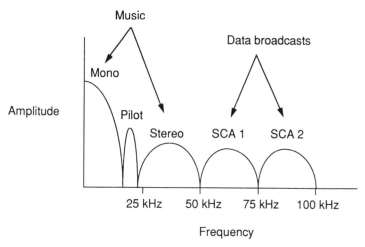

Figure 11.1 Subcarrier data (SCA 1 and SCA 2) rides piggyback on commercial FM broadcasts.

Seiko MessageWatch). Seiko is also working with Delco Electronics on an IVHS test in Seattle, demonstrating the delivery of traffic information to a car stereo.

Local roadside transmitters could serve as simple location beacons, or to warn of local hazards. Several technologies have been suggested, ranging from coded patterns of buried magnets (serving as location identifiers) to infrared transmitters. Local roadside voice transmitters operating at 1,610 kHz, a frequency that most AM radios can pick up, have become fairly prevalent in the United States for broadcasting traffic announcements.

Both existing and new two-way mobile radio systems could be used. SMR and particularly cellular telephone are attractive because of their large installed base. A number of SVRSs have been devised to exploit the cellular telephones and security alarms already found in many higher priced automobiles. Not surprisingly, Motorola is among those using existing two-way mobile communication services in conjunction with IVHS pilot programs.

New two-way mobile radio systems have been designed especially for vehicle-to-vehicle communications. The 60-GHz radio band is of great interest due to the presence of an oxygen absorption peak that limits transmitters (and potential interferers) to extremely short ranges. (As a result of O_2 absorption, atmospheric attenuation is 1,000 times greater at 60 GHz than at 15 GHz.) But most 60-GHz technology is still very expensive, having been developed by the U.S. military. Vehicular radar would operate at even higher frequencies. General Motors has petitioned the FCC to allocate 76 to 77 GHz to vehicular collision-avoidance systems.

IVHS Technologies a San Diego–based startup firm, has developed the VORAD™ (Vehicular Onboard Radio) "radar-based collision mitigation system." This is a small, low-power device that warns drivers (audible tone or synthesized

Figure 11.2 A package consisting of Cue Paging's DataPlatform radio modem, ComCard software, and Sharp Wizard are used to deliver TrafficAlert messages.

voice) of when their vehicle begins approaching another vehicle or object too rapidly. VORAD also enables what the firm calls "adaptive cruise control"—matching the speed of one's vehicle to that of the next vehicle ahead. The 0.5-mW radar signal operates at 24.125 GHz. In addition to using the radar for sensing and measuring, IVHS Technologies has experimented with piggyback data transmissions over the radar beam for vehicle-to-vehicle communications.

Local roadside transceivers are being used in navigation and other applications. For example, emergency vehicles can actuate traffic signals to halt cross-traffic as they approach an intersection. Siemens Automotive has developed a family of roadside and vehicular infrared transceivers for navigation purposes. A dash-mounted display (called the ALI-SCOUT) tells the driver when to turn and change lanes, and continually updates the estimated time of arrival. The ALI-SCOUT and associated infrared transceivers are being used in the FAST-TRAC (faster and safer travel through traffic routing and advanced controls) system in Oakland County, Michigan—a partnership between the Michigan Department of Transportation, the Road Commission for Oakland County, the FHA, and academic and industrial partners (Figures 11.3 and 11.4).

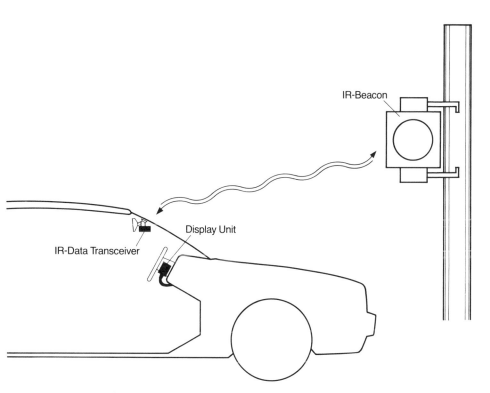

Figure 11.3 Siemens' infrared transceivers provide vehicle-to-road communications.

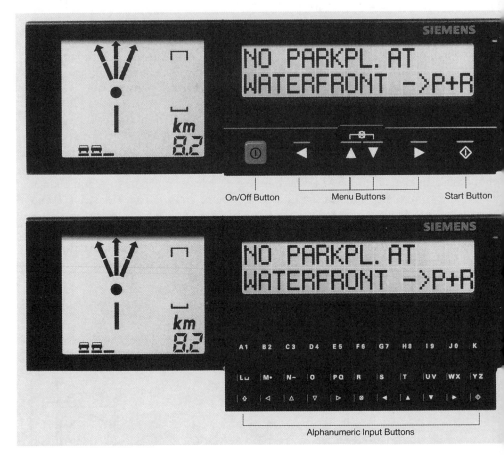

Figure 11.4 Siemens' ALI-SCOUT provides personal navigation in a compact package.

Japan has been testing the use of *leaky coax* installed along roadways, with road-to-car and car-to-road communications in the 1.5- and 2.6-GHz bands. One reason for using leaky coax is that 25% of the planned Tokyo-to-Kobe expressway will pass through tunnels to minimize real estate costs and environmental impact. Japan, by the way, is also experimenting with what it calls the GuideLight System: light sources on the roadway are activated as cars approach (yielding a savings over continuous overhead lighting) and can even be used to direct cars to change lanes (perhaps to avoid an accident site or stalled vehicle).

The use of cellular digital packet data (CDPD) for IVHS was demonstrated by AT&T, NavTech, and GTE Personal Communications Services at an IVHS America exhibition in April 1994. Remote control of a simulated variable message sign (VMS) was the featured application. The point these vendors were trying to make was, why develop a specialized IVHS communications infrastructure when commer-

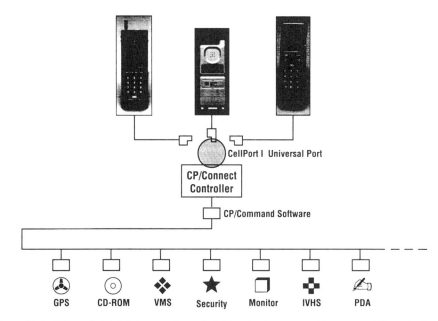

Figure 11.5 CellPort Labs' C/P Connect™ is a LAN for the fast-approaching automobile of the future.

cial services are already emerging that provide the necessary throughput and coverage?

CellPort Labs (Boulder, Colorado) is developing a LAN for vehicles that will enable different devices within an automobile to talk to each other (Figure 11.5). For example, the activation of an air bag could be used to instruct a cell phone to automatically dial (911) emergency service and report the vehicle's current location as provided by an onboard GPS receiver. Several cellular carriers have announced their support for CellPort's technology.

11.5 BRINGING IVHS TO MARKET

Making roads intelligent will not be an inexpensive proposition. IVHS America estimates over $200 billion will be spent on IVHS over the next 20 years in the United States. Of that amount, about $40 billion will go towards infrastructure, while nearly $170 billion will be spent by consumers for equipment and services.

But who will pay to build the $40 billion infrastructure? To some, the creation of an IVHS is the next moon shot—a project that can only be tackled by government. IVHS is about long-term goals; industry thinks primarily in terms of short-term objectives. IVHS seeks benefits for all of society, such as public safety, a better environment, and national competitiveness. Private firms seek profits for themselves.

Like the Apollo moon mission, IVHS could increase employment, stimulate technology development, and ultimately help create whole new industries.

Besides the increased tax burden, government involvement raises a number of thorny issues. For example, if a government-run platoon driving system takes control of your steering wheel as you merge onto a smart highway, who assumes accident liability? Experience suggests the legislature will simply pass a new law, exempting government from such obligations. In a more sinister future, could government use IVHS to track our movements? (Would this not be immensely useful to law enforcement officials conducting criminal investigations?) Perhaps the solution is to award local franchises for deploying, maintaining, and operating roadway sensors and communication systems.

Indeed, most proponents believe IVHS will succeed through a partnership between public agencies and private industry. Federal funding for state highways might be linked to smart-road initiatives. For better or worse, government is often the largest customer: IVHS technologies are already being employed in publicly owned commuter bus services and to automate highway toll collection. Like pollution controls, ATIS equipment could become standard equipment at an added cost of $800 to $1,200 per vehicle. Unfortunately, this might reduce congestion not by improving road utilization, but by making automobiles unaffordable to a large segment of society.

The alternative approach is to let private companies develop and market autonomous devices. Over 400,000 automobile navigation systems have already been sold in Japan. For example, Toshiba sells a system (for as much as $8,000) that includes two GPS receivers, a CD-ROM player (for electronic maps and associated databases), and a rearview camera.

General Motors would like to see navigation systems sell for about the same price as automotive air conditioners. Since the average consumer is sufficiently familiar with local streets, however, these systems are most likely to be sold to rental car companies and commercial fleets. General Motors' Oldsmobile division is selling a digital map-based system called GuideStar® that provides audible (synthesized voice) turn-by-turn directions; electronic yellow pages are a possible future enhancement. GuideStar was developed by Zexel USA in conjunction with Delco Electronics and Oldsmobile and retails for $1,995. Avis Rent A Car in San Jose, California, is equipping a fleet of cars with the system (Figure 11.6).

But the rental car business is fiercely competitive. It is difficult to justify loading cars with exotic extras unless there is a quick and certain payback. Such extras, for example, complicate fleet management as the rental car company must inventory vehicles both with and without navigation. Depending on the availability of other options (particularly if another rental return is late), the rental car company may be forced to upgrade a customer to a car with navigation at no extra charge.

Commercial fleets are more likely to make large up-front investments in order to reduce fuel waste, save time, and ensure safety. In many cases, they have already installed two-way radios. Onboard navigation can be employed to enhance cus-

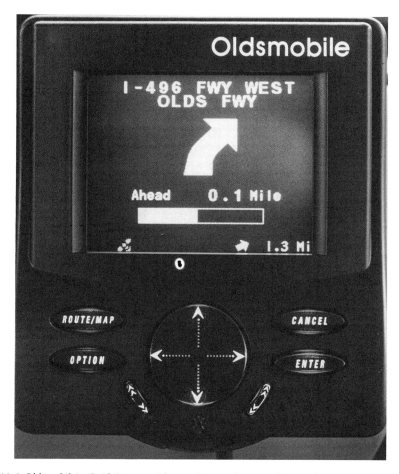

Figure 11.6 Oldsmobile's GuideStar provides audio turn-by-turn driving directions and supporting graphics.

tomer service (for example, by getting to the customer sooner). But widespread investment in technologies with long payback cycles vanished with the 1980s; in the 1990s, most companies demand payback within 18 to 24 months.

11.6 RF IDENTIFICATION

While IVHS planners dream up laundry lists of futuristic services, a number of application-specific technologies have emerged from the private sector and made their way to market. Used for years to track freight containers, RF/ID has more recently been adapted to automatic toll collection.

RF/ID accounts for a small piece of the burgeoning automatic identification industry. The largest and best known segment is laser-based bar-code scanning: grocery, hardware, clothing, and other retailers use bar-code scanning to track inventory and automate POS operations. RF/ID technology, in the form of antitheft tags, is becoming widely used in clothing stores. An alarm is triggered when someone tries to exit the store carrying a piece of clothing with its antitheft tag still attached; the tags are easily removed by sales clerks with special tools.

RF/ID systems not only detect presence, but also read stored data by bouncing RF beams off tags. RF/ID tags have several advantages over printed bar codes. Bar codes may only be read within line of sight, do not work well on fast moving objects, and are easily damaged or counterfeited. RF/ID tags, in contrast, can be read while attached to moving objects. Low-frequency tags operate between 100 and 400 kHz and have a range of 5 to 15 feet. High-frequency tags operate in the 900-MHz band and above, with a range of 40 to 150 feet.

RF/ID systems consist of two essential components: the base station reader and the RF/ID tag. RF/ID tags are actually portable programmable semiconductor memory devices. Different types of RF/ID tags are available featuring a variety of storage capacities (from 1 to 256,000 bits) and reading ranges. While battery-powered active tags offer the greatest capacity and range, they are presently outnumbered by batteryless passive tags, primarily used to thwart shoplifters and identify employees (security tags).

RF/ID tags may also be classified as read-only and read/write. Read-only tags are preprogrammed by the manufacturer and are used to assign identification numbers to objects, animals, and people. Read/write tags accept additional data or changes, and can be reprogrammed with a special terminal or base station. For example, Ford Motor Company's Louisville, Kentucky, Ranger and Bronco truck plants use RF/ID tags to carry special instructions forward to the next production stage.

Automated toll collection may prove to be the ultimate prize for RF/ID base station and tag manufacturers. Consumers could use read/write tags as debit cards (prepaid up to a fixed dollar amount). Each time the driver passes through the automated toll collection gate, the toll is deducted by overwriting the remaining balance. One idea under study in Europe is to link toll charges to traffic conditions. When roads are heavily congested, drivers would be charged more. Perhaps this rather simple technology is all that is needed to reduce congestion. On the other hand, we may be shocked to learn how much the average consumer is willing to pay to go home early.

11.7 STOLEN-VEHICLE RECOVERY SERVICE

Car thefts have reached epidemic proportions in the United States with over 1 million automobiles stolen per year at a cost of more than $8 billion annually.

Stolen-vehicle recovery technology began with the "bumper beeper"—a remotely activated hidden radio beacon. Today, there are several different SVRS providers, each using somewhat different technology. Some systems must be operated by law enforcement agencies (e.g., Lojack). Other systems are independently operated, provide 24-hour-per-day manned control centers, and assist police in tracking down stolen vehicles.

Here is a typical scenario. Once a car is reported stolen, the command to activate the hidden transmitter is broadcast over a wide area. In other words, a wireless, two-way data communications system is used. The now-operational beacon can be located through a process of triangulation, in which multiple receivers identify and compare the direction from which the signal emanates.

Although the stolen-vehicle recovery business has not met growth expectations, a few vendors recognized that they possess the technology, licenses, and infrastructure to provide other types of metropolitan area networking services, particularly those that require a combination of two-way mobile data and radiolocation. Some of these systems are now focusing on fleet management and IVHS-related applications.

11.8 PIONEERS OF AUTOMATED VEHICLE MONITORING

AirTouch Teletrac is a joint venture between AirTouch Corporation and North American Teletrac that builds and operates AVM networks. The firm currently operates in the 902-MHz band in Los Angeles, Chicago, Detroit, Dallas/Ft. Worth, Miami, and Houston and holds FCC licenses to build similar networks in approximately 140 MSAs across the United States.

AirTouch Teletrac is allied with equipment manufacturer Tadiran (Israel) and geographic information systems vendor Etak (Menlo Park, CA). AirTouch Teletrac plans to initiate service in New York, Washington/Baltimore, San Francisco, and San Diego next, and cover a total of 20 cities by 1996. Existing services include commercial-vehicle location, stolen-vehicle recovery, and roadside assistance. The company reports that public sector customers—such as transit, public safety, and public utilities—account for 40% of the vehicles on its network.

AirTouch Teletrac introduced its Fleet Director fleet management system in March 1993. Fleet Director combines AVL, mobile data communications, and vehicle security. An optional message display terminal (Coded Communications' CMX-1000 and CMX-4500) enables dispatchers to send messages directly to drivers. The target cost per user is about $30 per month. Fleet Director software (approximately $2,000) provides a digitized map of the AirTouch Teletrac (metropolitan) coverage area.

Teletrac uses RF spread spectrum telemetry over an 8-MHz slice of the 900-MHz band. The system can locate a vehicle to within ±100 feet. Teletrac's SVRS is activated when a car is started without the owner's key or when a standard auto alarm is tripped. A series of radio towers (e.g., 40 throughout the greater Los Angeles area) continuously monitor these "silent alarms" (Figure 11.7).

AirTouch Teletrac sees IVHSs, the mobile office, and personal communications as key growth areas. But there are more urgent applications: the firm's system will be installed in 200 Avis automobiles available at the Miami airport. (The State of Florida—which depends heavily on revenue from tourism—has suffered a nightmare in which several foreigners have been murdered on its roads.) At the push of an emergency button, the driver will be able to summon police—without having to provide directions. An acknowledgment LED will let the customer know the police are on their way. An AirTouch Teletrac operator will call the police and give them a description of the vehicle—make, model, color, license plate number, current location, and even speed and direction of travel.

Pinpoint Communications is a privately held firm founded in 1990 that has developed a metropolitan area mobile data/radiolocation network called ARRAY. The firm uses off-the-shelf technology to provide services ranging from roadside assistance to fleet management. Pinpoint hopes to construct a nationwide network operating in the 902- to 928-MHz AVM band (licensed under FCC Rules Part 90.239) in 40 to 50 of the largest U.S. cities. The firm introduced its technology in early 1993 in Washington, D.C.

Pinpoint uses a high-speed single channel to serve a large number of users—potentially up to one million per city. The mobile TransModem is expected to sell for about $300. Basic service will cost $10 to $15 per month (2,000 position fixes included), and data transmission will cost $0.01 per 20-character packet.

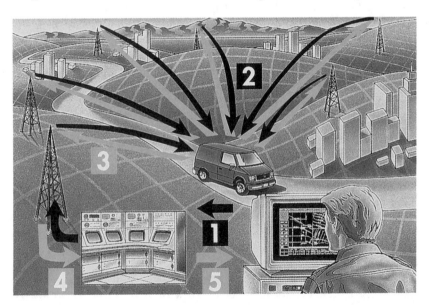

Figure 11.7 AirTouch Teletrac's Fleet Director™ helps automate dispatching and recover stolen vehicles.

Pinpoint's strategic partners include SFA, Inc., who will develop and manufacture the TransModem and the Intelligent Mobile Data Network (IMDN) ARRAY base stations—what Pinpoint describes as a "land-based antenna grid." ARRAY can pinpoint vehicles to within ±50 feet. The TransModem provides mobile data transmission at a sustained throughput of up to 38,400 bps with about 200-ms latency—very short by packet radio standards.

ARRAY sends data piggyback over spread spectrum ranging pulses. The instantaneous RF link rate is 380,000 bps. The ARRAY network determines the TransModem's location solely from data transmissions by measuring differences in time of arrival of mobile signals at pairs of base stations. ARRAY uses proprietary technology to keep all elements in the network time-synchronized, to provide communications path redundancy, and to overcome multipath propagation (Figure 11.8).

The firm sees IVHS and ultimately consumer applications for its technology. Pinpoint has demonstrated an Electronic Copilot—using Fujitsu's menu- and pen-based PoqetPad hardware platform—for the advanced traveler information system (ATIS). The menu includes roadside services (e.g., assistance with flat tires), traffic alerts, and electronic yellow pages. Pinpoint has also teamed up with Navigation Technologies to integrate electronic maps with its applications.

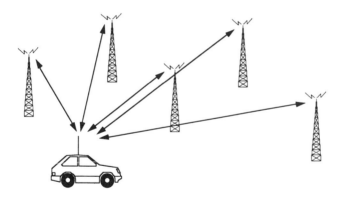

Mobile ID	Base #	Time
0019	001	10:17:43.35443
0019	002	10:17:43.35444
0019	003	10:17:43.35443
0019	004	10:17:43.35447
0019	005	10:17:43.35448

Figure 11.8 Pinpoint Communications' ARRAY network measures and compares the time of arrival of signals at 4 to 12 base stations to determine a vehicle's exact location.

11.9 ROAD WARRIORS

We live in an increasingly mobile society. There are more than 187 million registered vehicles in the United States alone. Mobile communications is important not only because it permits people to reclaim travel time for other purposes, but because it can help make the trip shorter, safer, and more pleasant.

But how will smart cars and smart roads succeed? The assumption that IVHS is the next moon shot and therefore government must play the lead role is simply not true. From navigation to fleet management, there are a multitude of compelling business and consumer applications. And there is no shortage of competing technologies, either. The first item on the agenda is to achieve the price/performance ratio necessary to entice customers.

While roads are rigid and hierarchical, intelligent road networks need not be. Instead of building IVHS from the top down—like a mainframe computer network—we should encourage it to grow from the bottom up—like personal computer networks. Certainly, local communities should be free to promote enhanced services or restrict the number of infrastructure providers as they see fit. But intelligent road networks will require, first and foremost, the support of business users and consumers.

Armed with the right products and services, we will get directions, check travel times, find and reserve parking spaces (perhaps the killer application in some cities?), and access local businesses. There will also be benefits to society: we will avoid congested roads, identify and compare public transportation alternatives, and access emergency services. The opportunities for intelligent road networks—plied by PDAs and personal communicators—are nothing short of fantastic.

SELECT BIBLIOGRAPHY

[1]   Ames, R., *Perspectives on Radio Frequency Identification*, New York: Van Nostrand Reinhold, 1990.
[2]   Catling, I., ed., *Advanced Technology for Road Transport IVHS and ATT*, Norwood, MA: Artech House, 1994.
[3]   Collier, W. C., and R. J. Weiland, "Smart Cars, Smart Highways," SEI Information Technology, *IEEE Spectrum*, April 1994.
[4]   Jurgen, R. K., "Smart Cars and Highways Go Global," *IEEE Spectrum*, May 1991.
[5]   Walker, J., ed., *Mobile Information Systems*, Norwood, MA: Artech House, 1990.

CHAPTER 12
▼▼▼

INTERACTIVE TELEVISION
AND BEYOND

12.1 DIGITAL CONVERGENCE?

Digital technology promises to revolutionize both the medium and the message. Telephone companies want to get into the cable TV business. Cable TV companies want to get into the telephone business. Computer companies want to manufacture set-top boxes—transforming hundreds of millions of living room televisions into interactive TVs (ITV). Even Hollywood wants to get into the act, abandoning traditional film in favor of digital multimedia content.

But most of the digital convergence talk is misplaced. Everyone assumes the old guard will create the new industry. Nothing could be further from the truth. Just as leading vacuum tube manufacturers did not rise to the challenge posed by semiconductors, and mainframe and minicomputer manufacturers failed to lead the personal computer revolution, it will be up to the next Intel or Microsoft to usher in the brave new world of interactive multimedia. The giants of the computer and telecommunication industries will leave the business of innovation and risk-taking to entrepreneurs—many of whom recognize wireless technology's potential to quickly fashion interactive multimedia networks out of today's telephone and television infrastructure.

ITV promises new forms of entertainment, education, shopping, and even political expression (so-called *teledemocracy*). Viewers will be able to compete with others while playing video games, order pizzas, search multimedia libraries, and register their opinions during talk shows. Digital audio broadcasting (DAB), advanced television (ATV), and data broadcasting will also contribute to the metamorphosis of broadcasting into "narrowcasting." The airwaves will be filled with personalized "infotainment."

Personal communicators will become the wireless remotes of the ITV age. For example, Interactive Network, Inc., has built its ITV strategy around a wireless information appliance. Eon Corporation has petitioned the FCC for permission to use the interactive video data service (IVDS)—created as an ITV enabler—for consumer messaging. And Sony's Magic Link personal communicator doubles as a remote control for the firm's home entertainment products.

12.2 INTERACTIVE TV

For better or worse, television has become the center of family entertainment. In the United States, over 235 million people live in homes with televisions. They are turned on an average of 7 hours and 14 minutes per day. Over 73 million video games and 64 million VCRs have also been sold. When it was first introduced, TV was heralded as an invention that would bring culture to the ordinary living room. Over the years, viewers have shown an unmistakable preference for soap operas, situation comedies, and endless sporting events. In 1967, FCC commissioner Lowinger called television a ". . . low-brow medium" and predicted all attempts to improve it would prove futile.

Today, many believe television *can* be significantly improved. The problem with today's television is that the viewer ("couch potato") is passive. New technology will enable the viewer to participate. The television will become a window into vast electronic libraries. Electronic shopping will help reduce road congestion and pollution. Electronic town hall meetings will usher in a new era of participatory democracy.

Others believe TV viewers are only interested in better picture and sound quality. Televisions and computers are entirely different products serving profoundly different markets. The PC is typically viewed from a distance of 12 to 18 inches and primarily used for work; the TV is viewed from a distance of 8 to 10 feet and primarily used for entertainment. HDTV sets will incorporate computers, but only to improve the picture. Desktop computers will receive and display video not for entertainment—but for business.

Is there a "killer application" for ITV? Conservative business planners are betting on enhanced TV services such as video on demand (VOD), interactive video games, and an electronic channel guide that automatically programs your VCR. Others believe the future is in home banking and merchandising. But there are problems with each of these applications.

Pay-per-view, reverse time shifting, and VOD simply enhance the (passive) viewing experience. Pay-per-view has already enjoyed some success, particularly in hotels. Reverse time shifting permits the viewer to schedule programs at their own convenience—competing with the distasteful process of programming one's own VCR. VOD requires sufficient bandwidth to download a feature-length movie in a few minutes—what some observers refer to as a *virtual VCR*.

In the United States, television has always been advertiser-driven. Perhaps ITV is a tool for advertisers, not viewers. Although it raises privacy concerns, ITV can be used to track individual viewing and purchasing habits. ITV may also be a new conduit to impulse buyers. ITV proponent Eon, for example, is working with Domino's Pizza and retailer J.C. Penney. Perhaps advertisers are the ones who must pay for—or at least subsidize—ITV.

Others have a more futuristic vision of ITV. Social commentator Marshall McLuhan predicted that one day we will talk back to our TV sets. Viewers will be able to control camera angles during live events, access indepth reports on subtopics during news or documentary shows, and even decide which version of a play or movie to watch. No longer passive, viewers will track down criminals during mystery shows and call the plays during football games. Well, perhaps high school football games.

A ubiquitous ITV infrastructure will open new doors for teleconferencing and *telepresence*. Perhaps the television, rather than the telephone, will serve as a platform for the elusive videophone service. Telepresence—a sort of networked virtual reality—will allow surgeons to operate remotely. Players will be able to fully immerse themselves in video games. But the big money will, no doubt, be made in *telesex*.

Cost, time, and lack of interest are factors that could turn out to be ITV show stoppers. Many homes already have a video game player, VCR, and two or more TVs. About 60% of U.S. households now subscribe to cable TV. Digital DBS services are coming on line. The problem: too many businesses are competing for consumers' discretionary spending money.

ITV hardware is already selling at home entertainment market prices. For example, Interactive Network sells its control unit for $199. ITV subscription charges are reasonable; basic service starts around $15 per month. But new ventures tend to believe people will buy if only the price is low enough. But price is not the only consideration.

Some analysts believe most consumers have simply exhausted their free time. Just as the rise of network TV in the early 1950s contributed to the decline of general-interest magazines, people will have to give up something in order to use ITV. When most people come home from work, the only machine they care to interact with is the refrigerator. Most TV viewers do not have the patience to manipulate a screen filled with menus, data, games, and services. But if this is true, ITV may succeed by emphasizing time-saving services: electronic shopping, home banking, and VOD.

12.2.1 ITV Technologies

A number of Interactive Television technologies are being developed, field tested, and deployed. The one thing common to all of them is the ability to broadcast data along with the TV program. Viewer responses may either be transmitted immediately (continuous ITV) or stored and sent in a batch transmission (end-of-program ITV). Let us examine the data transmission alternatives for ITV:

Plain Old Telephone Service

Dial-up is a ubiquitous, inexpensive, and popular medium for uploading viewer responses. The only drawbacks are that it requires a phone jack in the den or living room and, when used for ITV, the line is temporarily unavailable for voice calls. A system for broadcasting data to viewers during the TV program is also needed.

Two-Way Cable TV

Cable TV began life as community antenna TV (CATV). Its original purpose was to ensure quality reception for groups of homes in rural or fringe reception areas. Today, the United States has more than 11,000 cable systems serving over 57 million subscribers. Cable TV systems pass within reach of over 95% of U.S. homes.

Cable TV networks were originally designed as one-way broadcast systems. They are now upgrading to two-way transmission in anticipation of not only ITV, but also of competing with LECs in providing basic telephone services. Due to their tree topology and reliance on analog transmission, upstream communications have been a challenge, because the noise from each of the branches is additive. Digital channels solve this problem by regenerating the bits.

In 1972, the FCC mandated that all new cable TV systems had to support two-way communications. Warner Amex Cable Communications introduced its Qube two-way service using a low-speed return channel in 1977. Subscribers, however, proved apathetic to what were primarily merchandising services. The U.S. Supreme Court abolished the FCC's two-way requirement in 1979. Today, the major cable TV operators recognize the strategic importance of two-way communications.

Direct Broadcast Satellite

New DBS systems can reach homes equipped with relatively small and inexpensive antennas. (DBS should not to be confused with the large-dish antennas some households have used to intercept satellite distribution of video programming to local broadcast and cable TV operators; Congress outlawed unauthorized reception of scrambled satellite signals in 1984.)

Cable TV giant Telecommunications, Inc. (TCI) sees DBS as a means of reaching the estimated 10 to 12 million homes that currently lack access to cable TV service. Using digital technology, others want to compete head-on with terrestrial cable service. Outside the United States, DBS is the main platform for delivering HDTV. But the cost and complexity of a subscriber uplink may limit the use of DBS to the ITV downlink.

Multichannel Multipoint Distribution Service

Multichannel multipoint distribution service (MMDS) is better known as *wireless cable*. The FCC established MMDS in 1983, allocating frequencies in the 2.1- and 2.6-GHz microwave bands. Line-of-sight communication is required to small-dish antennas. Today, MMDS operators offer up to 33 channels of programming per community, in some cases by leasing channels from licensees in the adjacent Instructional Television Fixed Service (ITFS) and the private operational fixed services.

The growth of MMDS has been limited. Transmitters are restricted to a maximum output of 100W and the signals are easily blocked by buildings, hills, or foliage. But the main limitation has been channel capacity, preventing MMDS operators from offering the variety of programming found on cable-based systems.

Zenith Electronics (Glenview, Illinois) has developed a two-way interactive wireless cable system for applications such as pay-per-view. The subscriber uplink uses the ITFS audio response channel.

Low-Earth-Orbit Satellite

On the same day Vice President Albert Gore used the phrase "Global Information Infrastructure" instead of his oft-repeated term "National Information Infrastructure," Bill Gates and Craig McCaw announced their desire to construct a $9 billion, 840-satellite, interactive multimedia network operating in the 20- to 30-GHz Ka-band. The proposed Teledesic system would provide bandwidth on demand at rates from 16 kbps to 2.048 Mbps to users equipped with antennas as small as 16 cm in diameter (and up to 1.8m) and transmitting at power levels as little as .01W (up to a maximum of less than 5W). This system could serve ITV's two-way requirements, but will not be available until at least the year 2001 (see Chapter 9).

Mobile Data Networks

Mobile data services like ARDIS and RAM Mobile Data have expressed little interest in ITV. But several wireless ITV developers have shown keen interest in mobile data. Their systems must focus on covering residential areas, and they recognize that business users are also consumers. It is less likely, however, that nationwide mobile data networks will be used for ITV.

Interactive Video Data Services

IVDS was created specifically for ITV. The FCC allocated two 500-kHz licenses in the 218- to 219-MHz band per service area; each licensee will enjoy exclusive rights within a 40-mile radius. The first 18 licenses in 9 markets were awarded by lottery; the remaining 1,450 licenses in 725 markets were distributed via auctions bringing in nearly $250 million.

IVDS supports two-way data transmission. Base stations are permitted to run 1W to 20W effective radiated power (ERP)—depending on their proximity to channel 13 transmitters (which operate on an adjacent frequency). Home units are permitted a maximum ERP of 20W. Licenses are granted for 5 years and require 10% build-out per year, with 60% build-out completed within 5 years. The FCC has warned it will reclaim underused IVDS spectrum. Most licensees are expected to lease or purchase networks from companies such as Eon (discussed below).

FM Subcarrier Radio

ITV data may be broadcasted to viewers via FM subcarrier. Since FM subcarrier is one-way only, another technology (e.g., POTS) must be employed to upload viewers' responses.

Vertical Blanking Interval

When a television's electron scanning beam returns from the bottom right corner of the screen back to the top left corner to begin the next scanning sequence, the beam must be switched off. Data may be sent over the TV channel's frequency during this blanking interval without causing interference.

The transmission of text over the vertical blanking interval (VBI) was pioneered by the British Broadcasting Corporation (BBC) beginning in 1973 with its CEEFAX teletext service. (Independent British TV stations initiated a competing ORACLE service in 1975.) Teletext has been successful in England and on the European continent; it is a standard feature on most new TV sets, and an estimated 20% of British households have access to the service. In the United States, the PBS, NBC, and CBS television networks have all experimented with teletext. A service called SilentRadio broadcasts a single line of moving text and mosaic graphics intended for display in restaurants, bars, and banks. With the exception of closed captions for the hearing-impaired, however, VBI services have failed to win significant market support in the United States.

The National Television System Committee (NTSC) standard specifies 42 VBI lines per frame. Twenty lines are reserved for vertical synchronization, equalization pulses, other internal purposes, and closed captions for the hearing-impaired. This leaves 22 lines accessible to data broadcasters. Each VBI scan line can carry about 320 bits, or 40 ASCII text characters; there are 60 blanking intervals per sec. Therefore, a single scan line can support 2,400 characters per sec (24,000 bps). Using a simple decoder, the data may be overlaid on the TV picture or displayed separately.

ITV Applications

Some ITV players plan to use a combination of the services described above. For example, NTN Communications is using FM subcarrier, DBS, VBI, two-way cable TV, and POTS—depending on cost and availability in each geographic market.

12.2.2 Selected ITV Pioneers

Radio Telecom & Technology

Radio Telecom & Technology (RTT) (Riverside, California) has developed a wireless system called T-Net to facilitate rapid and cost-effective deployment of ITV. The firm, founded in 1984, describes T-Net as "a virtual cellular system" for both the ITV uplink and downlink.

T-Net was designed to transmit on an adjacent TV channel during the VBI—since adjacent channels are otherwise unoccupied. The central terminal system partitions the subscriber area into *virtual cells* based on precision timing. Instead of the traditional honeycomb cells, T-Net creates cells out of concentric circles subdivided by wedges. The concentric circles are defined by the time it takes the signals to propagate from base station to subscriber. RTT claims T-Net can serve up to 500,000 users within a 30-mile radius. T-Net runs at aggregate speeds in the millions of bits per second, the exact rate depending on network size (Figure 12.1).

Originally, RTT hoped to operate on adjacent channels in the TV bands. The firm pointed out that T-Net could support personal communications as well as ITV. The FCC rejected RTT's request in 1991 on the grounds that unused UHF channels were needed for advanced television. RTT countered that it was highly unlikely that every one of the 56 UHF channels would be consumed by the transition to HDTV, and in any event T-Net's spread spectrum–like signals could coexist with ATV. Subsequently, RTT modified T-Net to operate as a stand-alone system on dedicated IVDS channels.

The FCC relaxed the IVDS rules to accommodate T-Net. (RTT's persistent attempts to obtain spectrum for its innovative technology have made the firm some-

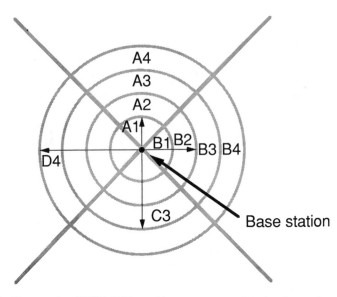

Figure 12.1 In this example of RTT's T-Net architecture, propagation time from a base station defines four concentric circles, which are further divided into four sectors, to create 16 cells.

thing of an industry martyr.) RTT is now focusing on selling its equipment to IVDS licensees and operating a backbone network. Japan's Oki Electric has agreed to manufacture T-Net set-top boxes, and the firms plan to begin delivering systems by the end of 1994.

It will cost an estimated $500,000 to $1 million per market to deploy T-Net. The system supports near video on demand (VOD), in which a movie starts every half hour, but not true VOD. Noting that pay-per-view currently represents less than 1% of all cable TV revenues, and that a large portion of VHS tape rentals are the 10 to 12 hottest movies, RTT believes that near VOD best fits established market needs.

Louis Martinez, RTT's president, sees a market composed of two major segments. First, there are basic interactive services for which the broadcaster provides the content and controls the screen—such as game shows, sports, and shopping. Second, there are transactional services for which a separate entity provides the content (and the viewer controls the screen)—such as banking, program guides, airline schedules, catalogs, and surveys. Martinez believes that basic interactive services will succeed first.

ITV will offer new capabilities to advertisers. RTT's system has the ability to identify which channel a viewer is watching—even if the viewer is not using interactive functions at the time. Advertisers will be able to offer electronic coupons; Hewlett-Packard is developing a set-top box printer for precisely this purpose. RTT is also promoting the idea of unsolicited response buttons using on-screen tomato and heart icons. If you do not like what you are seeing, press the tomato button; if you do like it, select the heart. The size of the displayed icon is determined by the audience's aggregate reaction.

RTT expects average monthly ITV revenue of $15 per month per household, consisting of $5 per month from each of three sources: the subscriber, advertisers, and service providers. RTT expects to sell its set-top receiver transmitter unit (RTU) for about $200. The RTU has four expansion slots for (1) data over video (DOV) encoder/decoder, (2) TV graphics card, (3) modem adapter card, and (4) home monitor card for security and other uses.

T-Net has been tested and demonstrated under experimental license in Salt Lake City (1985–1986) and Los Angeles (1987–1990). RTT also has an experimental license to test its system in conjunction with WNET in New York City on channel 13 (210 to 216 MHz) and the 1-MHz channel immediately above it (216 to 217 MHz).

Interactive Network, Inc.

Interactive Network (Mountain View, California) is a subscription-based ITV service that allows viewers to play along with TV programs using a wireless control unit. Service was initiated in Sacramento, San Francisco, and Chicago first, with 10 additional cities on the way. TCI is one of Interactive Network's major investors.

Data are broadcasted via FM subcarrier or the TV VBI to the control unit. The downlink runs 14.4 kbps, and the uplink runs 2,400 bps using dial-up access to a public (packet-switched) data network. The VBI transmission path is being provided by PBS National Datacast on a nationwide basis. When VBI is used, it is decoded by a set-top box and relayed to the handheld unit at 9,600 bps via an omnidirectional infrared link.

One surprising aspect of the system is that all of the ITV menus, prompts, and responses are displayed on Interactive Network's handheld control unit with hide-away keyboard. While the viewer must look at two different displays, the unit is not restricted to TV-based applications. The control unit is bundled with a 6-month subscription for $299. This includes the basic $15/month fee; if the viewer wants to compete for prizes, the monthly fee is bumped up to $25 (see Figure 12.2).

Interactive Network's CEO, David Lockton, firmly believes his company is developing the first consumer applications for mobile data. He believes the "killer application" is entertainment. Like RTT's Martinez, Lockton believes transaction services will have to be subsidized by advertisers.

Eon

Eon (Reston, Virginia) has developed a cellular IVDS network for two-way wireless communication. Eon, formerly known as TV Answer, has been an aggressive proponent of IVDS. The firm has formed alliances with Hewlett-Packard's ITV Appli-

Figure 12.2 Interactive Network's control unit enables viewers to play along with TV quiz shows.

ances Group (to manufacture home units selling for less than $500 each), General Instrument (to incorporate an Eon module in its set-top boxes), and Hughes Communications (to establish a national satellite network and information center).

The firm's revised network architecture incorporates remote sensors (up to 14 per cell site) to support low-power subscriber devices. VSAT earth stations will link the cell sites to the national information center. The system uses TDD and GPS for synchronization. Eon partitions each IVDS channel into 15 separate channels (to minimize interference between subscriber units), one of which is reserved for control functions. Like Interactive Network and RTT, Eon believes wireless offers the lowest cost per home—in fact, less than $2 per household. Wireless systems can also be deployed much faster than wired solutions.

Eon is working with dozens of potential service providers including PBS and certain grocery chains. PBS will use ITV for fund-raising and children's educational programs. Unlike Interactive Network, Eon uses translucent on-screen graphics. The control unit houses a mouseball and trigger. When the control unit is in "universal remote" mode, a keypad for functions like VCR programming will appear on the screen.

Eon does not see a single "killer application," but an array of personal services. Eon's program guide will automatically turn on the TV and/or VCR when a selected program airs. EON has announced agreements to develop ITV services with Capital Cities/ABC and NTN Communications targeting popular news programs, soap operas, and sporting events.

EON does not intend to obtain its own IVDS licenses. Instead, it will construct networks and lease them to licensed affiliates. The firm is ready for commercial roll-out. No monthly subscription fees are planned; either the viewer or an advertiser will pay EON for each use. (Eon, in turn, will pay a commission to the appropriate affiliate.)

Eon has petitioned the FCC to amend the IVDS Rules to permit the operation of mobile devices—what it calls *response transmitter units*—transmitting at 100-mW effective radiated power (ERP). In its petition, Eon states ". . . what will fuel the growth of the IVDS industry . . . is to provide useful and convenient consumer messaging." Parents will be able to send "come home now" messages to their children. Subscribers will be able to operate their VCRs remotely.

12.3 ADVANCED TELEVISION

ATV is part of what some see as a natural progression for television technology: from black and white (yesterday)—to color (today)—to high resolution (tomorrow)—to three-dimensional (distant future). But others say what television really needs is not an enhanced picture, but enhanced programming.

Nevertheless, ATV began as an effort to improve the picture. At first, the migration to digital transmission seemed far-fetched; most efforts concentrated on improved analog transmission. One of the first goals, in fact, was to eliminate annoying "ghost" pictures—a problem that does not even exist with digital transmission. Later, the goal expanded to creating wide-screen, motion-picture-quality pictures in the living room—(high definition TV (HDTV)).

The current analog standard in the United States (and several other countries) was developed by the NTSC in the 1940s. An NTSC frame is composed of 525 interlaced lines (i.e., each picture is painted in two scans). The original NTSC standard was enhanced to support color (remaining backwards-compatible with black and white) in the 1950s.

The Japanese were the first to pursue high-resolution TV. Japan's NHK proposed a system called MUSE HDTV in the early 1980s, employing more than twice as many interlaced scanning lines (1,125) per picture as current systems. Many feared Japan was too far ahead to be caught. The Europeans, however, came up with HD-MAC to thwart what appeared to be the U.S.-backed Japanese standard.

Both the European and Japanese HDTV systems were hybrids—they combined digital processing with analog transmission. The United States entered the game when several players decided to push ahead with all-digital solutions. Digital offers many advantages, both for transmitting and processing pictures. Digital pulses can be regenerated, FEC can be employed, signals can be stored and manipulated, spectrum efficiency can be improved, no alignment is required during manufacturing, and the electronics can be expected to drop in price as higher levels of silicon integration are achieved. But perhaps the biggest advantage of digital transmission is that viewers will have more control over their TVs—even the ability to control them with home computers. Once all TV programs are produced and distributed as digital bit streams, it will be possible to move freely from one medium to another, mixing and matching video clips, sound bytes, and text.

The U.S.'s commitment to digital transmission shook Europe and Japan. The European community, which had spent $750 million developing a satellite version of their analog HDTV system, canceled the project in the spring of 1993. Closing ranks against the Japanese, several European players decided to team up with U.S. firms. As part of a comeback strategy, Japan has begun work on its next-generation HDTV system, the all-digital ultradefinition TV (UDTV), which offers twice the resolution of HDTV and is delivered via satellite.

The picture is now clear: HDTV is at the center of a political battle between the major developed countries. At issue are independence from foreign suppliers, domination of the global market, and domestic high-tech jobs. But the strategic vision is now blurred: the United States seems determined to follow a model that has failed miserably overseas—the government-industry alliance. In the migration to HDTV, U.S. manufacturers hope to regain a large chunk of the domestic market for TV sets. One player even suggests that because larger screens will entail higher transportation costs, domestic suppliers will be at an advantage.

Once again, the spectrum question. In late 1990, the FCC announced it would select a simulcast standard for HDTV. The idea was to simply give the terrestrial HDTV market to existing broadcasters, permitting each to operate on two independent channels during a transition period (up to 15 years). The original plan was to select an ATV standard by mid-1993 with the help of the independent Advisory Committee on Advanced Television Service (ACATS).

Six systems were offered for testing. The FCC decided to select an all-digital system after General Instrument/MIT (two digital systems), Philips-Thomson-Sarnoff, and Zenith-AT&T announced their commitment to digital. But it was unable to identify a single system that was clearly better than the rest.

It was at this point the FCC decided to bail out. Rather than letting itself get embroiled in what would likely end in a court battle, the FCC suggested the remaining players pool their efforts and bring HDTV to market as quickly as possible. And so the "Grand Alliance" between Zenith, AT&T, General Instrument, MIT, Philips, Thomson, and SRI International's David Sarnoff Research Center was born.

The Grand Alliance proposed its first standards—encompassing compression, transport, scanning, audio, and transmission (described below)—to ACATS in May 1993. The standard reflects a strong commitment to compatibility between HDTV and computer video.

Compression.   Video compression techniques take into account the limitations of human vision to maximize compression while minimizing perceived picture quality loss. The Grand Alliance has settled on an approach similar to the Motion Picture

Experts Group's MPEG++ system (using 1,080 × 1,920 pixels). This approach has also been embraced by the computer industry.

Transport. A communication protocol featuring fixed-length packets has been proposed to support video, audio, text, graphics, and other types of data.

Scanning. Television has traditionally used interlaced scanning (each picture is painted in two passes), while computers have used progressive scanning (each line is scanned in sequence). The Grand Alliance will initially support both formats, but promises migration to the exclusive use of progressive scanning at 60 frames/sec. We can also expect HDTV to incorporate another element favored by the computer industry: square pixels.

Audio. Selected standards are Dolby AC-3, Musicam 5.1, and MIT-AC.

Transmission. Four competing systems still under evaluation: two variants of quadrature amplitude modulation (QAM) and two variants of vestigial sideband (VSB).

The head of MIT's Media Lab, Nicholas Negroponte, wants to go even further. Negroponte is calling for the development of open architecture TV (OATV)—a formal marriage between the television and computer. According to Negroponte, interlaced versus progressive scanning, number of scan lines, frame rate, aspect ratio, and pixel shape are all "nonissues" (Table 12.1). What keeps Negroponte up at night is fear that the United States, Europe, and Japan will develop different standards for digitizing program content.

Once again, we are confronted by experts who demand an emerging industry commit itself to one path and one path only. Pity the hapless consumer who buys a $4,000 HDTV set that cannot receive all high-definition broadcasts and eventually becomes obsolete! (Another view: the person who can afford, and chooses to purchase, a $4,000 television deserves no protection other than against fraud.)

Proponents of early HDTV standards point to AM stereo (stereo broadcasts in the 0.5- to 1.5-MHz AM broadcast band) as an example of the dire consequences that may follow in the absence of standards. In 1982, the FCC was in an antireg-

Table 12.1
Several Key Differences Between Today's Personal Computers and Televisions

	Personal Computer	*Television*
Viewing distance	12–18 inches	8–10 feet
Location in home	Home office	Living room
Scanning method	Progressive	Interlaced
Pixel shape	Square	Round
User input	Keyboard, mouse	Remote control
Primary use	Business	Entertainment

ulatory mood and decided to let the market choose an AM stereo standard. It began with five competing systems. More than a decade later, AM stereo has yet to achieve success.

But was the lack of a standard the reason AM stereo never caught on? Although AM stereo is intended not only to provide "right" and "left" channels, but also to improve audio fidelity, consumers have always associated AM radio with poor fidelity. The reason AM stereo failed was not lack of a standard, but lack of a market.

Nevertheless, the FCC has committed to selecting and mandating a terrestrial HDTV standard by 1995. Combine that with threats to take away broadcasters' NTSC channels and you have a real incentive to migrate to HDTV. Unfortunately, broadcast TV has lost a considerable portion of its market to cable TV. It is not clear the majority of TV stations can afford the upgrade to HDTV—let alone the cost of simulcasting during a transition period that is certain to last years.

According to one member of the Grand Alliance, the first HDTVs are expected to appear on the market by early 1997 at $3,500 to $6,000 each (for sets with 52- to 60-inch diagonal screens). So much for protecting ordinary consumers. Initially high prices are inevitable, but only competition can bring the dramatic reductions necessary for a mass market. One day, digital TV will become cheaper than analog TV. We will have entered the age of personal television—an age in which viewers will be able to control and manipulate programming.

12.4 DIGITAL AUDIO BROADCASTING

DAB promises to enhance urban, rural, and international radio broadcasting. In cities, terrestrial-based DAB will deliver CD-quality sound. In large countries like India and China, satellite-based DAB will provide reliable coverage for nationwide broadcast networks. Although currently unpopular in the United States, international broadcasting may stand to gain the most. DAB would enable broadcasters like the BBC and Voice of America (VOA) to deliver local-quality signals to all parts of the globe. The ability to access local broadcasts from any country on earth will be another benefit of the personal communications revolution.

Like HDTV, DAB suggests new links between FM radios and other microprocessor-based devices. Once audio is transmitted as a digital bit stream, there is no reason conventional radio programming cannot include text, paging, real-time traffic information, and more. But one thorny issue plagues the migration to DAB: combined with digital audio tape (DAT), it would enable consumers to make perfect recordings of music broadcast over the airwaves—something the recording industry does not want to see happen. Like most new technologies, DAB is a double-edged sword. But it is also inevitable, and attempts to block it will prove futile and counterproductive.

12.4.1 DAB Technologies

The United States was a late comer to the brave new world of DAB, as it was with HDTV. The first issue that had to be addressed was how DAB should be delivered. There are three proposed solutions: (1) out-of-band (i.e., outside the existing commercial FM broadcast band), (2) inband, off-channel (i.e., in the existing FM broadcast band, but on unused channels), and (3) inband, on-channel (DAB buried underneath standard FM broadcasts).

Not all DAB applications require CD-quality audio. The primary purpose of satellite DAB is to reach small, mobile receivers anywhere within a large geographic area. But receiving relatively high-speed satellite signals with mobile receivers is something of a challenge. High-gain satellite antennas capable of 50 dBW and greater EIRP are required. The ideal frequency range for satellite DAB is 1 to 3 GHz, where atmospheric attenuation is low (compared to higher frequencies), and receiver antennas are small (compared to lower frequencies).

By now the reader may suspect the existence of a DAB government-industry alliance. This one, at least, is built around a fairly interesting technology. Europe's Eureka 147 employs coded orthogonal frequency-division multiplexing (COFDM) to deliver up to 16 CD-quality stereo channels (plus data services) per 4-MHz slice of spectrum. A scaled-down version (1.5 MHz) has been proposed in the United States.

In COFDM, the bit streams from each of several broadcasters are interleaved over multiple narrowband carriers. In other words, the broadcasters share a composite wideband channel, each enjoying greater immunity to multipath fading. In fact, COFDM uses multipath signals constructively, supporting simultaneous terrestrial and satellite broadcasting on the same channel. Such a combination could offer truly continuous coverage, relying on terrestrial transmission for urban areas (where tall buildings obscure satellite coverage) and satellite transmission for rural areas.

The Electronic Industries Association (EIA) Digital Audio Radio Subcommittee is working on a DAB standard for the United States. More than 10 proposals have been submitted. Candidates for EIA-sponsored testing include three inband, on-channel; one inband, adjacent-channel; one Eureka (terrestrial plus satellite) L-band; and one satellite-only system. Each of these systems is expected to support auxiliary data.

12.4.2 Selected Pioneers of Digital Audio Broadcasting

Two of the largest players involved in the development of DAB are the VOA and AT&T. VOA proposes a 256-kbps system including FEC and auxiliary data and occupying 200- to 250-kHz channels. AT&T is interested in DAB as an application of its audio compression technology. AT&T's Perceptual Audio Coding (PAC) enables low transmission rates, without loss of perceived quality, by taking into account human hearing characteristics. Uncompressed high-fidelity audio requires about 1.4 Mbps; PAC does the job at 128 to 160 kbps—about a 10:1 compression ratio.

Kintel Technologies has dropped out of formal participation in EIA testing, but has developed a unique on-channel solution called Power Multiplexing® (PMX). This technology has ramifications well beyond DAB. PMX enables the decoding of two unequal strength carriers on the same frequency. If it works, PMX could nearly double today's spectrum capacity. It also has obvious military applications such as clandestine communications.

PMX works by inverting the stronger signal (creating a replica 180 deg out of phase with the original) and adding it back to the received signal. In essence, the stronger signal is subtracted out or attenuated enough to ensure capture of the weaker signal. The advantage of PMX in DAB is that it would permit a broadcaster to simply add a (weaker) digital signal to an existing analog signal. Digital receivers would employ PMX to filter out the stronger analog signal. Analog receivers would simply capture the stronger analog signal, as before.

12.5 DATA BROADCASTING

Digital TV and radio herald a new world of data broadcasting. No longer will broadcast programming be limited to music and talk. New services are already emerging for distributing text, photographs, sound clips, drawings, films, and more. The opportunities seem limitless, but so far data broadcasting has met only limited success. PDAs and personal communicators are about to change that.

Data broadcasting will be used to distribute time-sensitive material like software upgrades, updates to repair manuals, electronic newspapers and magazines, traffic information, and financial and scientific news. (In fact, some of these services already exist.) Broadcasting the information is not the main challenge; it is the development of an inexpensive tool for capturing, organizing, and presenting it. PDAs and personal communicators, equipped with software for learning the user's interests and filtering out all else, will enable users to pull the data they need out of the airwaves.

12.5.1 Data Broadcasting Technologies

Data broadcasting will eventually become a huge business. What we are talking about is nothing less than a dynamic, up-to-the-minute, library over the airwaves. For now, data broadcasting is primarily a telephone company bypass technology. In the future, users equipped with PDAs and personal communicators will pull down location and time-specific information for both business and personal use.

Paging was the first personal data broadcasting service. Nationwide carriers such as Cue Paging, Motorola's EMBARC, and Skytel are now adding broadcasts of news headlines, stock quotes, and traffic information to their repertoire. In addition to timely delivery of information to individuals, these services can compete with overnight couriers by, for example, offering near-instant electronic delivery of product updates and price lists to field sales personnel scattered across the country.

Data superimposed on TV broadcasts, in most cases using the VBI, are a major focus of data broadcast technology development. It is perhaps here that the wholesale broadcasting of software, books, and newspapers will take place. Data are also broadcast directly via satellite. In addition, there is interest in using broadband PCS for high-speed (T1 speeds and faster) data broadcasting.

The Radio Broadcast Data System (RBDS) is an effort to standardize and promote the transmission of auxiliary data over commercial FM radio stations using FM subcarrier technology. RBDS was endorsed by the National Radio Systems Committee (NRSC) in late 1993, runs 1,200 bps, and is based on Europe's Radio Data Standard (RDS). RBDS could be used to display frequency, station logo, and brief text messages on radios equipped with LCDs. In Europe, RDS is used to deliver real-time traffic information. Some proponents in the United States want to make data a mandatory feature on all automotive AM/FM radios. In one proposed (and chilling) scenario, emergency notification systems would be able to turn on receivers remotely to ensure that users receive safety alerts.

12.5.2 Data Broadcasting Pioneers

WavePhore, Inc.

WavePhore (Tempe, Arizona) believes data broadcasting will play a major role on the information superhighway. The firm's first product, the Video 7500, is a TV channel modem capable of receiving data at 384,000 bps. WavePhore's TVT1 technology enables digital data broadcasting on an active TV channel without employing (or interfering with) the VBI or audio subchannel.

According to WavePhore, TVT1 uses what amounts to blank space within today's broadcast TV channel. The data broadcast signal is specially shaped to prevent interference with the video luminance or chrominance signals and transmitted at about 10 dB above the video noise floor. TVT1 products are planned in three configurations: stand-alone (initially about $3,000), PC add-on board, and chip set. The system may be used over either broadcast or cable TV. WavePhore expects to increase its data rate to 1.5 Mbps—enough to carry one MPEG compressed video signal (VHS-quality) along with the standard NTSC signal.

One possible application of TVT1 is as a LAN extender. Data could be received from a LAN from anywhere within the broadcast zone. Although a phone line or cellular connection would be needed for the reverse channel, there are many LAN applications that are asymmetrical, such as downloading a database or text document.

Advanced Digital TeleCorporation

Advanced Digital TeleCorporation (ADTEC), a Brewster, New York–based division of Jeen International, has developed a system for broadcasting 160,000 pages per day using four VBI lines—what it calls a "broadband store-and-forward service." Information content may include VOD, video catalogs, music, games, product data, news, and financial data.

ADTEC's TeleData Recorder (TDR) is the result of a cooperative development effort with ITT Intermetal. The unit is about the size of a VCR and works with any TV or PC. It enables selective capturing, storing, and displaying of information. Up to 4 MB (1,000 pages) can be stored in the terminal unit, which handles text (in up to 32 different languages) and graphics. The first model, the TeleData 8000, is expected to sell for about $400.

Data broadcasts currently contain about 400% overhead to ensure data integrity. The user terminal profile can be remotely configured by the broadcaster. Broadcast input may be via satellite, VHF/UHF, or cable system. Each TDR is individually addressable and includes an uplink modem.

ADTEC claims its TDR offers "instant ITV" with (1) the economy of point-to-multipoint transmission, (2) the convenience of a TV display, (3) the ability to be operated by anyone, (4) immediate access for small-and medium-sized businesses to the information superhighway, and (5) access for advertisers (Figure 12.3).

Over 4,000 information categories are possible. The TDR has the ability to tune in to the appropriate frequency and time to capture information from a preselected category. ADTEC expects its terminal to be subsidized by information suppliers, with both subscriber-and supplier-paid information options. For example, Dow Jones news might be available at no charge with advertising, but the subscriber would have to pay for the news if the advertising was stripped out.

12.6 FROM BROADCASTING TO NARROWCASTING

Today's main broadcast technologies—TV, radio, and paging—are overdue for a change. The migration to digital presents new opportunities, but it also poses new challenges. Once broadcasting converts to a digital bit stream, it will be possible to intermix a wide variety of data. Figuring out what kinds of information people want, and how they will use them, is the hard part.

But one thing is clear. Up until now, the consumer could only receive TV and radio broadcasts. Digital broadcasting is not simply high-resolution video or CD-quality audio, it is software-controlled TV and radio. Both viewers and advertisers will gain unprecedented control. Each consumer will be able to put data broadcasts to individual use—from personalized electronic newspapers to TV systems that automatically learn a viewer's interests and record the appropriate programs.

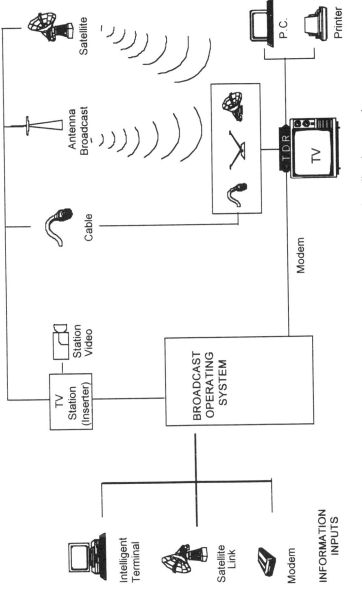

Figure 12.3 ADTEC will use the broadcast television infrastructure to create a "broadband store-and-forward service."

SELECT BIBLIOGRAPHY

[1]   Browning, J., "EuroTechnoPork," *Wired*, May/June 1993.

[2]   De Sonne, M., ed., *Multimedia 2000*, Washington, DC: National Association of Broadcasters, 1993.

[3]   De Sonne, M., ed., *Advanced Broadcast/Media Technologies*, Washington, DC: National Association of Broadcasters, 1992.

[4]   Interview with Gary Arlen, *Wireless Industry Prospectus*, Datacomm Research Company, 1993.

[5]   Lippman, A., "HDTV Sparks a Digital Revolution," *Byte*, December 1990.

[6]   Negroponte, N., "HDTV: What's Wrong With This Picture?" *Wired*, Premier Issue, 1993.

[7]   Negroponte, N., "Set-Top Box as Electronic Tollbooth: Why We Need Open Architecture TV," *Wired*, September/October 1993.

[8]   Cross, B., WavePhore, presentation at Wireless User Conference, Orlando, Florida, 1994.

[9]   Lupin, B., TV Answer, presentation at Wireless Data Conference, San Jose, California, 1992.

[10]   Martinez, L., Radio, Telecom & Technology, presentation at Wireless User Conference, Orlando, Florida, 1994.

[11]   Lockton, D., Interactive Networks, presentation at Wireless Data Conference, San Jose, California, 1993.

[12]   Tatsuno, S., *Created in Japan: From Imitators to World-Class Innovators*, New York: HarperBusiness, 1990.

CHAPTER 13
▼▼▼

WHY PERSONAL COMMUNICATIONS WILL SUCCEED

Voice communication over wire, although technically possible, hardly seems a reliable means of conversing, and if it were, it is highly unlikely that the public would avail themselves of such a service.

—*Boston Herald*, 1891

13.1 NOWHERE TO HIDE?

No, the *Boston Herald* was not prescient regarding wireless communications.

Whenever new technology appears there are always those who greet it with comments like "it won't work" and "even if it did, no one would want it." The pessimism expressed by the *Boston Herald* at the close of the last century has become all too common. We are told PDAs do not deliver what they promise, and even if they did, there is no market for them. People fear personal communications will leave them nowhere to hide. According to some industry analysts, mobile computing is more hype than reality.

It is always easier to criticize businesses built around new technologies than it is to champion them. This is why people admire successful inventors and entrepreneurs. They exhibit a faith in oneself and persistence that few can muster. If we were all entrepreneurs, most of us would be either rich or bankrupt; very few would land somewhere in between.

Today it is easy to laugh at the *Boston Herald*. Hindsight is always perfect. Those who examined the first telephones, however, saw a technology that was not ready for the mass market. The obstacles that stood in the path of Alexander Graham Bell were enormous. There was no existing infrastructure through which he could leverage his invention. It took a visionary, indeed, to realize that the first scratchy-sounding telephones—originally developed for the purpose of aiding the deaf—might evolve sufficiently to serve a few niche markets.

Personal communications is poised to change the world. But the "experts" see only what is presented to them. Journalists see vendors trying to convince customers to adopt technology that is expensive, difficult to use, and incompatible with existing systems. Consultants see minefields through which they must guide their clients. Entrenched players see markets to be protected. And politicians see technologies to be regulated and special interest groups to be served.

No one knows exactly how personal communications will succeed. Although technology often succeeds in unexpected ways, there are some interesting signposts. For example, most people assume mobility will prove to be personal communications' greatest gift. In solving the basic networking challenges posed by mobile communications, we will develop powerful features that will also benefit fixed users. *Mobile communications will be as successful at the desktop as it will be in the field.*

In the mobile environment, we must identify users whenever and wherever they appear, enable their subscription rights and apply any restrictions, locate them wherever they happen to go, and always bill their home network. More importantly, because mobile users may drift in and out of the network, we must learn to communicate in time slices.

Today's knowledge workers are deluged with e-mail, faxes, and telephone calls. Between meetings and the demands of the phone, many find it increasingly difficult to get any work done. What they need—what they are willing to pay for—is a technology that will enable them to reduce the time and effort they spend communicating. Users spend too much time listening to voice mail messages, calling people back and not reaching them, and—worst of all—getting ensnared in useless conversations when they make the mistake of personally answering the phone.

Personal communications will liberate us from dependence on real-time interactive communications. By firing off text messages, faxes, and digitized sound bytes, we will reach people and get answers more quickly. We will develop tools to filter out junk calls, junk e-mail, and junk faxes. Rather than disturbing every individual on the planet, telemarketers will learn to narrowcast their messages to those who truly have a need or interest. (In fact, we may not have to pay for personal communication services as advertisers offer to pick up the tab in exchange for information about our interests, needs, and habits.)

Imagine sitting in a meeting with your electronic notepad. Suddenly, the "inbox" icon begins to flash. You tap on it, and a hand-scratched note from a colleague in another city appears, with a short but rather urgent question. During a brief lull in the meeting, you scratch out the answer and tap the "send" icon. Had your colleague been forced to reach you by telephone or fax, it might have taken 24 hours or more to get the answer, or you might have been forced to leave an important meeting.

Personal communications is not about computing and communicating anywhere, anytime. To most users, that would be a nightmare. What they want is access to information and services relevant to their environment and time. Few people will gladly field unsolicited phone calls while relaxing on the beach; but they will want to access the latest weather forecast, information about nearby restaurants, and directions back to their parked car. *Location-specific information services, rather than the nebulous benefits of communicating anytime, anywhere, hold the key to success for personal communications.*

13.2 INFORMATION BOULEVARDS, SIDE STREETS, AND BACK ALLEYS

One of the assumptions we have challenged is the dependence on narrowband radio engendered by the "spectrum shortage." There are two prerequisites to the success of personal communications: (1) governments must permit competition in all radio services, as well as permit radio services to freely compete with wireline services, and (2) governments must allow new broadband radio technologies to be introduced into the radio spectrum.

The notion that radio is an inherently narrowband medium is false. The electromagnetic spectrum is the same in free space as it is inside cable. The primary limitation is interference, not bandwidth. More importantly, narrowband communication is unacceptable to most users, who expect wireless networks to deliver the same or better performance than wire-based networks. We can begin to understand why mobile data service has enjoyed so little success when we consider that today's mobile data networks only handle short messages, yet charge several times as much as landline networks. Users comprehend what vendors have failed to grasp: wireless communication requires less infrastructure than wire-based communication and should cost less to use. Mobile data will not succeed until vendors construct networks supporting higher throughputs and greater capacity.

Unfortunately, today's wireless data services were defined in terms of the past rather than the future. The first, primitive mobile data applications operated at extremely low speeds and, consequently, were only good for exchanging very short messages. It is now assumed that most mobile data applications will entail brief, intermittent transactions. In reality, business users will want to send and receive faxes while on the road; access desktop PCs, LAN servers, and host computers remotely; exchange multimedia documents with colleagues who are also mobile; and perform data-intensive operations like downloading digitized maps from wireless information services.

There is much talk about the information superhighway. If roadways are an apt analogy, we should note that most drivers spend the majority of their time on local streets. It is not that their interests are strictly parochial; it is just that they require the convenience and variety of many roads leading to nearby places, rather than a superhighway linking major hubs.

In fact, the idea that the United States needs an information superhighway seems totally misplaced. By most measures, the United States already has several information superhighways. What the United States needs now is a greater selection of "on-ramps." And this is where personal communications can play an important role. Wireless technology is the key to competition in the local loop. Users are not clamoring for the highly centralized infrastructure suggested by the information superhighway; they are asking for more choices. Personal communications will give users access to information boulevards, side streets, and back alleys—where the real action will take place.

The main goal of the national information infrastructure, as laid out by Vice President Albert Gore, is to ensure that society does not become segregated into information haves and have-nots. But government intervention carries its own risks. In the past, the government has granted protected markets (i.e., monopolies) as a quid pro quo for universal access. Today, government seems especially preoccupied with preventing (or at least limiting) the privacy made easy by digital transmission. While it is not clear that government can ensure there will be no information have-nots—after all, some people assiduously avoid public libraries—it is clear that personal communications will enable users to bypass the local access monopolies created and maintained by government.

13.3 WINNERS AND LOSERS

Everyone wants to know who will be the winners in the personal communications revolution. Industry analysts point to the entrenched giants who appear to hold a lock on the financial resources, engineering talent, manufacturing capacity, and distribution channels.

Daily the press focuses on the efforts of the largest players as they jockey for position in the nascent personal communications market. As the demand for services like paging and cellular telephone continue to escalate, and while government officials talk about deregulation and broader competition, a reshuffling of long-distance and local-access players is under way. AT&T has merged with McCaw Cellular, Bell Atlantic and NYNEX want to combine their cellular operations, and MCI had planned to invest in Nextel, but pulled back at the last minute. AT&T insists it is not trying to get back into the local telephone business, but it is certainly trying to bypass LEC monopolies; local-access charges account for an estimated 45% of long-distance costs. The Department of Justice approved the AT&T-McCaw merger with several caveats, one of which was that AT&T and McCaw should not be permitted to bundle cellular and long-distance service. In an attempt to protect consumers, the government is denying them lower prices.

Nevertheless, we should not be fooled into believing the only remaining task is for the established leaders to divide up the personal communications market. When semiconductors emerged, it was assumed that the new industry would be dominated by the leading vacuum tube manufacturers. Many expected IBM, DEC, Burroughs, and Wang would grab the lion's share of the new PC business. More often than not, entrenched players carry too much baggage when they attempt to enter new markets. Fearful of cannibalizing their existing businesses, they attempt to subjugate rather than exploit new technologies. *It is as hard, if not harder, for an entrenched player to succeed in a new market as it is for a startup firm.*

While governments around the world attempt to kick-start domestic industries, they would do better to study the PC revolution. The PC blossomed in a highly unregulated environment. If the government helped accelerate the development of microprocessors, it was primarily as the largest consumer of electronic systems used in defense and the exploration of outer space. But PCs succeeded in ways that surprised even industry pioneers. Arguably, the lack of early standards helped rather than hindered the PC. Both open (e.g., IBM-compatible) and proprietary (e.g., Macintosh) platforms have prospered. The best thing government can do for the personal communications industry is reallocate large tracts of spectrum—and then get out of the way.

13.4 THE GENIE IS OUT OF THE BOTTLE

The success of personal communications is inevitable. Just as large corporations were unable to stem the flow of PCs, the "authorities" will be unable to hold back the personal communications revolution. Everywhere, rigid and hierarchical networks are giving way to networks that are flat and fluid. Personal communications will take things the next step—ushering in the age of spontaneous and ad hoc networking.

To understand how personal communications will succeed, we need to think in terms of what might motivate people to stop saying, "Why would I want it?" and start saying "How did I ever get by without it?" Personal communications will succeed because it will establish new standards for timely access to people and information. Today, the measure of how smart you are is not how much you know, but how much you know that is timely.

Like PCs, personal communicators will tap the power of silicon and gallium arsenide, becoming increasingly powerful while rapidly declining in price. Just as today's desktop PC is more powerful than the mainframe of 20 years ago, the personal communicator of the next decade will be able to do more than today's telephones, fax machines, and LANs.

Although technology advances incrementally, the gains are cumulative. But few if any of us can see around the many corners that lie ahead. The best way to ensure personal communications will succeed is to permit it to evolve in as many different directions as entrepreneurs and investors care to pursue.

SELECT BIBLIOGRAPHY

[1]   Kirby, R. S., S. Withington, A. B. Darling, and F. G. Kilgour, *Engineering in History*, New York: Dover Publications, 1990.
[2]   Calhoun, G., *Wireless Access and the Local Telephone Network*, Norwood, MA: Artech House, 1992.

APPENDIX
▼▼▼

DIRECTORY OF COMPANIES AND ORGANIZATIONS MENTIONED IN THIS BOOK

Accessline Technologies, Inc.
11201 SE 8th Street
Bellevue, WA 98004
Tel: 203 781–9056

Advanced Digital TeleCorporation
Orchard Ridge Building, Fields Lane
Brewster, NY 10509
Tel: 914 277–1012; Fax: 914 277–1015

Air Communications
274 San Geronimo Way
Sunnyvale, CA 94086
Tel: 408 749–9883; Fax: 408 749–8089

AirTouch Teletrac
7391 Lincoln Way
Garden Grove, CA 92641
Tel: 714 897–0877; Fax: 310 338–7199

Note: Addresses and phone numbers are subject to change.

American Mobile Satellite Corporation
10802 Parkridge Boulevard
Reston, VA 22091
Tel: 703 758–6000; Fax: 703 758–6239

Ameritech Cellular Services
2000 West Ameritech Center Drive
Hoffman Estates, IL 60195–5000
Tel: 708 706–7600

ARDIS
300 Knightsbridge Parkway
Lincolnshire, IL 60069
Tel: 708 913–4330; Fax: 708 913–4340

AT&T Paradyne
8545 126th Avenue North
Largo, FL 34649–2836
Tel: 813 530–8638; Fax: 813 532–5949

AT&T PersonaLink
400 Interface Parkway
Parsippany, NJ 07054
Tel: 201 331–4132; Fax: 201 331–4505

BellSouth Cellular Corporation
1100 Peachtree Street NE, 12th Floor
Atlanta, GA 30309–4599
Tel: 404 249–2000; Fax: 404 249–0304

Casio Personal Communications
570 Mount Pleasant Avenue
Dover, NJ 07801
Tel: 201 361–5400; Fax: 201 361–3819

CDPD Forum
401 North Michigan Avenue, Suite 2200
Chicago, IL 60611–4267
Tel: 1 800 335–2373

CellPort Labs, Inc.
4730 Walnut Street
Boulder, CO 80301
Tel: 303 541–0722; Fax: 303 541–0731

Clinicom
4720 Walnut Street, Suite 106
Boulder, CO 80301
Tel: 303 443–9660; Fax: 303 442–4916

Communications Intelligence Corporation
275 Shoreline Drive
Redwood Shores, CA 94065–1413
Tel: 415 802–7888; Fax: 415 802–7777

Comsat Mobile Communications Corporation
950 L'Enfant Plaza SW
Washington, D.C. 20024
Tel: 202 863–6746; Fax: 202 488–3814

Cue Paging Corporation
2737 Campus Drive
Irvine, CA 92715
Tel: 714 752–9200; Fax: 714 833–9318

Dauphin Technology
377 East Butterfield Road, Suite 900
Lombard, IL 60148
Tel: 708 971–3400; Fax: 708 971–8553

Destineer Corporation
200 South Lamar Street
Jackson, MS 39201
Tel: 601 944–7484; Fax: 601 944–7415

Eon Corporation
1941 Roland Clarke Place
Reston, VA 22091–1405
Tel: 703 715–8600; Fax: 703 715–8806

Ericsson GE Mobile Data
45 C Commerce Way
Totowa, NJ 07512
Tel: 201 890–3600; Fax: 201 256–8768

Ex Machina
45 East 89th Street, Suite #39-A
New York, NY 10128–1251
Tel: 212 843–0000; Fax: 212 843–0029

Federal Communications Commission
1919 M Street NW, Room 814
Washington, D.C. 20554
Tel: 202 632–6600; Fax: 202 632–0163

General Magic
2465 Latham Street
Mountain View, CA 94040
Tel: 415 965–0400; Fax: 415 965–9424

Geotek Communications
20 Craig Road
Montvale, NJ 07445
Tel: 201 930–9305; Fax: 201 930–9614

GeoWorks
960 Atlantic Avenue
Alameda, CA 94501
Tel: 510 814–1660; Fax: 510 814–4250

Hughes Network Systems
11717 Exploration Lane
Germantown, MD 20876
Tel: 301 428–7165; Fax: 301 428–2801

Institute of Electrical and Electronics Engineers (IEEE)
345 East 47th Street
New York, NY 10017
Tel: 212 705–7555; Fax: 212 705–7453

Infrared Data Association
23 San Marino Court
Walnut Creek, CA 94598
Tel: 510 943–6546; Fax: 510 934–5241

INMARSAT
40 Melton Street, Euston Square
London NW1 2EQ
United Kingdom
Tel: 011 441 387–9089

Intelligent Transportation Society of America
400 Virginia Avenue SW, Suite 800
Washington, D.C. 20024
Tel: 202 484–4847; Fax: 202 484–3483

Interactive Network
1991 Landings Drive
Mountain View, CA 94043
Tel: 415 960–1000; Fax: 408 324–2001

InterDigital Communications
2200 Renaissance Boulevard, Suite 105
King of Prussia, PA 19406
Tel: 215 278–7800; Fax: 215 278–6801

Iridium, Inc.
1401 H Street NW
Washington, D.C. 20005
Tel: 202 326–5600; Fax: 202 842–0006

IVHS Technologies
10802 Willow Court
San Diego, CA 92127
Tel: 619 674–1200

Kintel Technologies, Inc.
P.O. Box 32550
San Jose, CA 95152
Tel: 408 729–3838; Fax: 408 926–1003

McCaw Cellular Communications
5400 Carillon Point
Kirkland, WA 98033
Tel: 206 827–4500; Fax: 206 828–8616

Meteor Communications Corporation
6020 South 190th Street
Kent, WA 98032
Tel: 206 251–9411

Metricom
980 University Avenue
Los Gatos, CA 95030–2375
Tel: 408 399–8200; Fax: 408 399–8274

Microcom
500 River Ridge Drive
Norwood, MA 02062–5028
Tel: 617 551–1688

Motorola EMBARC
1500 Gateway Boulevard
Boynton Beach, FL 33426–8292
Tel: 407 364–2000; Fax: 407 364–3683

Motorola, Inc.
1303 East Algonquin Road
Schaumburg, IL 60196
Tel: 708 576–5000

National Association of Broadcasters
1771 N Street NW
Washington, D.C. 20036–2891
Tel: 202 429–5376

Navigation Technologies
740 East Arques Avenue
Sunnyvale, CA 94086
Tel: 408 736–3700; Fax: 408 736–3734

Nextel
201 Route 17 North
Rutherford, NJ 07070
Tel: 201 438–1400

Norand
550 2nd Street SE
Cedar Rapids, IA 52401
Tel: 319 369–3100; Fax: 319 369–3299

National Telecommunications and Information Administration
14th & Constitution Avenue NW, Room 4898
Washington, D.C. 20230
Tel: 202 377–1850

Official Airline Guides
2000 Clearwater Drive
Oak Brook, IL 60521
Tel: 708 574–6000; Fax: 708 574–6075

Omnipoint Corporation
7150 Campus Drive
Colorado Springs, CO 80920
Tel: 719 548–1200; Fax: 719 548–1393

Orbital Communications Corporation
21700 Atlantic Boulevard
Dulles, VA 20166
Tel: 703 406–5300; Fax: 703 406–3508

Paging Network, Inc.
4965 Preston Park Boulevard, Suite 600
Plano, TX 75093
Tel: 214 985–4100; Fax: 214 985–6717

Palm Computing
4410 El Camino Real, Suite 108
Los Altos, CA 94022
Tel: 415 949–9560; Fax: 415 949–0147

Panasonic Communications
Two Panasonic Way
Secaucus, NJ 07094
Tel: 201 348–7000; Fax: 201 392–6023

ParaGraph International
1309 South Mary Avenue, Suite 150
Sunnyvale, CA 94087
Tel: 408 522–3000; Fax: 408 746–2813

PenWare
845 Page Mill Road
Palo Alto, CA 94304–1011
Tel: 415 858–4920; Fax: 415 858–4929

Personal Communications Industry Association
1019 19th Street NW, Suite 1100
Washington, D.C. 20036
Tel: 202 467–4770; Fax: 202 467–6987

Personal Computer Memory Card International Association
1030 B East Duane Avenue
Sunnyvale, CA 94086
Tel: 408 720–0107

Photonics Corporation
2940 North 1st Street
San Jose, CA 95134
Tel: 408 955–7930; Fax: 408 955–7950

Pinpoint Communications, Inc.
12750 Merit Drive, Suite 800, Park Central VII
Dallas, TX 75251
Tel: 214 789–8900; Fax: 214 789–8989

Proxim
295 North Bernardo Avenue
Mountain View, CA 94043
Tel: 415 960–1630; Fax: 415 960–1984

Qualcomm
6455 Lusk Boulevard
San Diego, CA 92121
Tel: 619 587–1121; Fax: 619 587–8276

Racotek
7301 Ohms Lane, Suite 200
Minneapolis, MN 55439
Tel: 612 832–9800; Fax: 612 832–9383

Radio Telecom & Technology, Inc.
6951 Flight Road, Suite 210
Riverside, CA 92504
Tel: 909 687–3660; Fax: 909 687–3892

RadioMail Corporation
2600 Campus Drive
San Mateo, CA 94403
Tel: 415 572–6001; Fax: 415 322–1753

Radish Communications
5744 Central Avenue
Boulder, CO 80308–3220
Tel: 303 443–2237; Fax: 303 443–1659

RAM Mobile Data
3 University Plaza, Suite 600
Hackensack, NJ 07601
Tel: 201 343–9400; Fax: 201 343–8795

Reflection Technology
230 2nd Avenue
Waltham, MA 02154
Tel: 617 890–5905; Fax: 617 890–5918

Roadshow International
8300 Greensboro Drive, Suite 400
McLean, VA 22102-3604
Tel: 703 356-9797

Seiko Communications
1625 NW AmberGlen Court, Suite 140
Beaverton, OR 97006
Tel: 503 531-3450; Fax: 503 531-1550

Siemens Automotive LP
2400 Executive Hills Drive
Auburn Hills, MI 48326-2980
Tel: 810 253-2642; Fax: 810 253-2998

SkyTel
1350 I Street NW, Suite 1100
Washington, D.C. 20005
Tel: 202 408-7444/1 800 543-5053

Sony Corporation
1 Sony Drive
Park Ridge, NJ 07656
Tel: 201 930-7834; Fax: 201 358-4058

SpectraLink
1650 38th Street, Suite 202E
Boulder, CO 80301
Tel: 303 440-5330

Spectrix Corporation
906 University Place
Evanston, IL 60201-3121
Tel: 708 491-4534; Fax: 708 467-1094

Stanford Telecom
480 Java Drive
Sunnyvale, CA 94089-1125
Tel: 408 745-0818; Fax: 408 541-9030

Symbol Technologies
116 Wilbur Place
Bohemia, NY 11716
Tel: 516 563-2400; Fax: 516 244-4645

Telecommunications Industry Association (TIA)
2001 Pennsylvania Avenue NW, Suite 800
Washington, D.C. 20006
Tel: 202 457–4912; Fax: 202 457–4939

Telular
920 Deerfield Parkway
Buffalo Grove, IL 60089
Tel: 708 256–8000; Fax: 708 465–4501

Telxon
3330 West Market Street
Akron, OH 44313–3352
Tel: 216 867–3700

Tetherless Access
43730 Vista Del Mar
Fremont, CA 94539–6250
Tel: 510 659–0809; Fax: 510 770–9854

Trimble Navigation
645 North Mary Avenue
Sunnyvale, CA 94088
Tel: 408 730–2900; Fax: 408 737–6057

U.S. Government Printing Office
Superintendent of Documents
Washington, D.C. 20402–9371
Tel: 202 783–3238

WavePhore, Inc.
2601 West Broadway Road
Tempe, AZ 85282
Tel: 602 438–8700; Fax: 602 438–8890

Zenith Data Systems
2150 East Lake Cook Road
Buffalo Grove, IL 60089
Tel: 708 808–5000; Fax: 708 808–4377

▼▼▼

GLOSSARY

ACATS	Advisory Committee on Advanced Television Service
ACTS	Advanced Communications Technology Satellite
AIN	advanced intelligent network
AM	amplitude modulation
AMPS	advanced mobile phone system
AMSC	American Mobile Satellite Corporation
AMTICS	advanced mobile traffic information and communication system
ANSI	American National Standards Institute
APC	American Personal Communications
APTS	advanced public transportation systems
ARDIS	advanced radio data information service
ARQ	automatic repeat query
ASCII	American National Standard Code for Information Interchange
ASIC	application-specific integrated circuit
ATG	air-to-ground service
ATIS	Advanced Traveler Information Systems
ATMS	Advanced Traffic Management Systems
ATT	advanced transport telematics
ATV	advanced television
AVC	automatic vehicle classification
AVCS	advanced vehicle control systems
AVL	automatic vehicle locating

AVM	automatic vehicle monitoring
B-CDMA	broadband CDMA
BAMS	Bell Atlantic Mobile Systems
BBC	British Broadcasting Corporation
BER	bit error rate
BETRS	Basic Exchange Telecommunications Radio Service
BTA	basic trading area
C/A	coarse acquisition
CAD	computer-aided dispatching
CAI	common air interface
CARIN	Car Information and Navigation System
CARS	cable television relay service
CATV	community antenna TV
CCD	consumer computing device
CCIR	International Radio Consultative Committee
CCITT	Consultative Committee for International Telegraph and Telephone
CDCS	continuous dynamic channel selection
CDLC	Cellular Data Link Control
CDMA	code-division multiple access
CDPD	cellular digital packet data
CEC	commission of the European Communities
CEPT	Committee of European Posts and Telecommunications administrations
CMRS	commercial mobile radio service
COFDM	coded orthogonal frequency-division multiplexing
Comsat	Communications Satellite Corporation
CPE	customer premises equipment
CRC	cyclic redundancy check
CRISP	C-Language Rational Instruction Set Processor
CT1	first-generation cordless telephone
CT2	second-generation cordless telephone
CT2Plus	second-generation cordless phone, enhanced
CT3	third-generation cordless telephone
CTIA	Cellular Telephone Industry Association
CTM	cellular telephone modem
CUG	closed user group
CVO	commercial vehicle operation
CW	continuous wave
D-AMPS	digital AMPS
DAB	digital audio broadcasting

DAT	digital audio tape
dB	decibel, a logarithmic measure of relative signal strength
DBS	direct broadcast satellite
DCA	dynamic channel allocation
DCS-1800	Digital Cellular System for 1,800 MHz
DCT	digital cordless telephone
DECT	digital European cordless technology
DES	Data Encryption Standard
DGPS	differential GPS
DIAD	delivery information acquisition device
DMS	digital messaging service
DOC	Department of Communications (Canada)
DOT	U.S. Department of Transportation
DRG	Direction de Reglementation Generale
DRIVE	Dedicated Road Infrastructure for Vehicle safety
DS-SS	direct sequence spread spectrum
DSI	digital speech interpolation
DSMA	data sense multiple access
DSMR	digital SMR
DSP	digital signal processor
DTI	Department of Trade and Industry (U.K.)
DTMF	dual-tone multifrequency
DTS	Digital Termination Service
e-mail	electronic mail
E-TDMA	extended TDMA
ECSA	Exchange Carriers Standards Association
EDA	embedded document architecture
EIA	Electronic Industries Association
EMA	Electronic Mail Association
EMBARC	Electronic Mail Broadcast to a Roaming Computer
ERMES	European radio message system
ERP	effective radiated power
ESMR	enhanced SMR
ETA	estimated time of arrival
ETC	Enhanced Throughput Cellular
ETSI	European Telecommunications Standards Institute
FAA	Federal Aviation Administration
FAST	frequency agile sharing technology
FAST-TRAC	faster and safer travel through traffic routing and advanced controls
FCC	Federal Communications Commission
FCC	forward control channel

FDD	frequency-division duplexing
FDMA	frequency-division multiple access
FEC	forward error correction
FH-SS	frequency-hopping spread spectrum
FHA	Federal Highway Administration
FHMA	frequency-hopping multiple access
FM	frequency modulation
FPLMTS	future public land mobile telecommunications service
FSK	frequency shift keying
FVC	forward voice channel
FWA	fixed wireless access
GEOS	geosynchronous-earth-orbit satellite
GII	global information infrastructure
GIS	geographical information system
GMSK	Gaussian minimum-shift keying
GPS	global positioning system
GSM	global system for mobile communication
HAZMAT	hazardous material
HDTV	high-definition television
HEOS	high-elliptical-orbit satellite
HF	high-frequency
HIPERLAN	High-Performance European Radio LAN
HIS	hospital information system
HP	Hewlett-Packard
HPCN	hybrid personal communications network
I/O	input/output
IC	integrated circuit
IDOT	Illinois Department of Transportation
IEC	International Electrotechnical Commission
IEEE	Institute of Electrical and Electronics Engineers
IMDN	Intelligent Mobile Data Network
IMTS	improved mobile telephone service
Inmarsat	International Maritime Satellite Organization
Intelsat	International Telecommunications Satellite Organization
IrDA	Infrared Data Association
IS	Interim Standard
ISI	intersymbol interference
ISM	industrial, scientific, and medical bands
ISO	International Organization for Standardization

ISTEA	Intermodal Surface Transportation Efficiency Act
ISV	independent software vendor
ITFS	Instructional Television Fixed Service
ITS	intelligent transportation systems
ITU	International Telecommunications Union
ITV	interactive television
IVDS	interactive video data service
IVHS	intelligent vehicle highway system
IXC	interexchange carrier
JTC	Joint Technical Committee on Wireless Access
KTS	key telephone system
LAN	local-area network
LCD	liquid crystal display
LEC	local exchange carrier
LED	light-emitting diode
LEOS	low-earth-orbit satellite
LMDS	local multipoint distribution service
LORAN	long-range aid to navigation
LOS	line of sight
MAC	media access control
MAS	multiple address system
MCC	Mobile Communications Controller
MCU	master control unit
MIN	mobile ID number
MIRS	Motorola Integrated Radio System
MMDS	multichannel multipoint distribution service
MNA	mobile navigation assistant
MNP	Microcom Networking Protocol
MoU	Memorandum of Understanding
MPEG	Motion Picture Experts Group
MSA	metropolitan statistical area
MSAT	mobile satellite
MSS	mobile satellite service
MTA	major trading area
MTBF	mean time before failure
MTEL	Mobile Telecommunications Technologies, Inc.
MTS	mobile telephone service
MTSO	mobile telephone switching office

N-AMPS	narrowband AMPS
N-CDMA	narrowband CDMA
NAB	National Association of Broadcasters
NAM	numeric assignment module
NASA	National Aeronautics and Space Administration
NCC	network control center
NDRC	National Defense Research Council
NIC	network interface card
NII	national information infrastructure
NOAA	National Oceanic and Atmospheric Administration
NOI	Notice of Inquiry
NPRM	Notice of Proposed Rulemaking
NRSC	National Radio Systems Committee
NTIA	National Telecommunications and Information Administration
NTSC	National Television System Committee
NWN	nationwide wireless network

OAG	Official Airlines Guide
OATV	open architecture TV
OSI	Open Systems Interconnection

P code	precision code
PAC	Perceptual Audio Coding
PageNet	Paging Network, Inc.
PBX	private branch exchange
PCIA	Personal Communications Industry Association
PCM	pulse-coded modulation
PCMCIA	Personal Computer Memory Card International Association
PCN	personal communications network
PCS	personal communications service
PDA	personal digital assistant
PIA	personal information appliance
PIM	personal information manager
PIN	personal identification number
PIP	personal information processor
PM	phase modulation
PMRS	private mobile radio service
PN	pseudorandom noise
POCSAG	Post Office Code Standardization Advisory Group
POFMS	Private Operational Fixed Microwave Service
POS	point of sale
POTS	plain old telephone service

PROMETHEUS Program for European Traffic with Highest Efficiency and
Unprecedented Safety
PSK   phase-shift keying
PSTN   public switched telephone network
PTN   personal telecommunications number
PTT   post, telephone, and telegraph
PTT   push to talk
PUC   public utility commission

QAM   quadrature amplitude modulation

RACS   radio/automobile communication system
RAM   random-access memory
RBDS   Radio Broadcast Data System
RBOC   Regional Bell Operating Companies
RCC   radio common carrier
RCC   reverse control channel
RCU   remote cell unit
RDS   Radio Data Standard
RF/DC   radio frequency/data collection
RF/ID   radio frequency/identification
RISC   reduced instruction set computer
RJ-11   Registered Jack 11, U.S. standard modular phone jack
RLP 1   Radio Link Protocol 1
ROM   read-only memory
RPE-LPC   regular pulse excited, long-term prediction
RSA   rural service area; rural statistical area
RTI   road transport informatics
RTU   receiver transmitter unit
RVC   reverse voice channel

SA   selective availability
SAT   supervisory audio tone
SDMA   space-division multiple access
SI   systems integrator
SIM   subscriber identity module
SIR   Serial infrared
SMR   specialized mobile radio
SMS   short-message service
SNA   Systems Network Architecture
SNR   signal-to-noise ratio

SOCRATES	System of Cellular Radio for Traffic Efficiency and Safety
SS#7	Signaling System No. 7
ST	signaling tone
SVRS	stolen-vehicle recovery service
T1 (DS1)	1.544-Mbps digital channel
T3 (DS3)	45-Mbps digital channel
TCI	Telecommunications, Inc.
TCP/IP	Transmission Control Protocol/Internet Protocol
TDD	time-division duplexing
TDMA	time-division multiple access
TETRA	trans-European trunked radio
TIA	Telecommunications Industry Association
UCT	universal coordinated time
UDPCS	universal digital personal communication system
UDTV	ultradefinition TV
UHF	ultrahigh-frequency radio band
UPCS	unlicensed personal communications service
UPS	United Parcel Service
UTAM	Unlicensed PCS Ad Hoc Committee for 2 GHz Microwave Transition and Management
VAD	voice activity detection
VAR	value-added reseller
VBI	vertical blanking interval
VHF	very-high-frequency radio band
VICS	vehicle information communication system
VLSI	very-large-scale integration
VMS	variable message sign
VOA	Voice of America
VOD	video on demand
VORAD	Vehicular Onboard Radio
VSAT	very-small-aperture terminal
VSB	vestigial sideband
VSELP	vector-sum excited linear predictive coding
WACS	wireless access communications systems
WARC	World Administrative Radio Conference
WIM	weigh-in-motion
WIN	wireless inbuilding network
WINForum	Wireless Information Networks Forum
WKTS	wireless key telephone system

WLAN	wireless local-area network
WOSA	Windows Open Services Architecture
WPBX	wireless PBX

X.25	CCITT standard for packet-switched networks
X.400	CCITT standard for interconnection of electronic mail
X.500	CCITT standard for directory services used by X.400-compliant networks
XIP	execute in place

10BASE2	Ethernet standard for thinNet (thin coax)
10BASE5	Ethernet standard for thick coax
10BASE-T	Ethernet standard for twisted-pair wiring

▼▼▼

ABOUT THE AUTHOR

Ira Brodsky is President of Datacomm Research Company, a Wilmette, Illinois-based management consulting firm specializing in wireless computing and telecommunications. Mr. Brodsky is an extensively published author and sought after speaker. He is Chairman of the "Wireless Data Conference and Exposition" and has served on the Advisory Board for COMDEX, the U.S.'s largest computer trade show. His articles have appeared in leading trade publications such as *Network World, Data Communications, Telecommunications,* and *Business Communications Review.* Mr. Brodsky is frequently quoted in the general and business press, including *The Wall Street Journal, The New York Times, Business Week, Fortune, Information Week, Computer World,* and *Telecommunications Reports*

Prior to founding Datacomm Research Company, Mr. Brodsky held positions with U.S. Robotics, Tektronix, and Gandalf Technologies. Mr. Brodsky received his B.A. in Philosophy from Northwestern University (Evanston, Illinois). He is a member of the IEEE. He may be reached via electronic mail at *brodsky@radiomail.net* or *5196661@mcimail.com*

▼▼▼

INDEX

The Artech House Telecommunications Library

Vinton G. Cerf, Series Editor

For further information on these and other Artech House titles, contact:

Artech House	Artech House
685 Canton Street	Portland House, Stag Place
Norwood, MA 02062	London SW1E 5XA England
617-769-9750	+44 (0) 171-973-8077
Fax: 617-769-6334	Fax: +44 (0) 171-630-0166
Telex: 951-659	Telex: 951-659
email: artech@world.std.com	email: bookco@artech.demon.co.uk